SPACE SCIENCE, EXPLORATION AND POLICIES

GALAXIES

EVOLUTION, MORPHOLOGY AND DYNAMICS

Space Science, Exploration and Policies

Additional books in this series can be found on Nova's website
under the Series tab.

Additional e-books in this series can be found on Nova's website
under the e-book tab.

SPACE SCIENCE, EXPLORATION AND POLICIES

GALAXIES

EVOLUTION, MORPHOLOGY AND DYNAMICS

ABRAN GUERRERA SANCHEZ
AND
RUTILIO E. DELGADO MARQUEZ
EDITORS

nova publishers
New York

For permission to use material from this book please contact us:
Telephone 631-231-7269; Fax 631-231-8175
Web Site: http://www.novapublishers.com

LIBRARY OF CONGRESS CATALOGING-IN-PUBLICATION DATA

Galaxies : evolution, morphology, and dynamics / editors, Abran Guerrera Sanchez and Rutilio E. Delgado Marquez.
p. cm.
Includes bibliographical references and index.
ISBN 978-1-62081-185-6 (hardcover)
1. Galaxies. I. Sanchez, Abran Guerrera. II. Marquez, Rutilio E. Delgado.
QB857.G393 2011
523.1'12--dc23

2012007371

Published by Nova Science Publishers, Inc. † New York

CONTENTS

PREFACE

In this book, the authors present current research in galactic study including its evolution, morphology and dynamics. Topics included in this compilation include the nature of motion in quiet and active galaxies with a satellite companion; empirical age-metallicity relation and empirical metallicity distribution of long-lived stars of different populations; radio quiet AGN properties vs. spin paradigm; and turbulent formation of protogalaxies at the end of the plasma epoch.

While the existence of intrinsic galaxy alignments have been repeatedly detected in a number of surveys, evidence results from two novel findings have been added that come primarily from several weak lensing studies which aims to quantify the gravitationally-induced shear field that distorts the background galaxy images. In addition, this allows for X-ray searches on substructure and the study of the dynamical state of galaxy clusters and their surroundings. Chapter 1 provides an outline for the observational and theoretical evidences of the formation of coherent large-scale cosmic structures; *i.e.*, structures showing the presence of either a non-random distribution of spin vectors of neighboring galaxies, or a correlation in their morphology, or simply non-random orientations of major axes of neighboring galaxies and clusters within filaments and superclusters.

The authors have already done an extensive discussion on the observational evidence for non-random orientations of major axes in neighboring galaxies and clusters within filaments and superclusters. The origin of such a coherence might be encoded into the nonlinear evolution of the density field through tidal interactions. But also, as the authors argue here, it could already be imprinted, back in time in the primordial structure of Gaussian density fields, as a consequence of a quantifiable cross-talk from small to large scales. It will be shown that if any coherent length exists, it would mainly be determined by the primordial tidal shear, especially from that produced by larger structures. In this picture, interclusters filaments of galaxies are enhanced by the non-linear evolution of the density field, with residual substructures keeping memory of their anisotropic origin. The authors finally discuss in Chapter 2, the possibility of a common, primordial origin for the wide gamut of alignment effects.

In Chapter 3, the authors present a galactic gravitational model of three degrees of freedom, in order to study and reveal the character of the orbits of the stars, in a binary stellar system composed of a primary quiet or active galaxy and a small satellite companion galaxy. The authors' main dynamical analysis will be focused on the behavior of the primary galaxy. They investigate in detail the regular or chaotic nature of motion, in two different cases: (i) the time independent model in both 2D and 3D dynamical systems and (ii) the time evolving 3D model.

For the description of the structure of the 2D system, the authors used the classical method of the Poincaré $x - p_x$, $y = 0$, $p_y > 0$ phase plane. In order to study the structure of the phase space of the 3D system, they take sections in the plane $y = 0$ of the 3D orbits, whose initial conditions differ from the plane parent periodic orbits, only by the z component. The set of the 4-dimensional points in the (x, p_x, z, p_z) phase space are projected on the (z, p_z) plane. The maximum Lyapunov Characteristic Exponent (L.C.E) is used in order to make an estimation of the chaoticity of this galactic system, in both 2D and 3D dynamical models. In order to distinguish between ordered and chaotic motion in the 2D system, the authors apply the $S(g)$ dynamical spectrum, while for the 3D Hamiltonian system they use a more advanced spectral definition, that is the $S(w)$ dynamical spectrum

The authors' numerical calculations indicate that the percentage of the chaotic orbits increases when the primary galaxy has a dense and massive nucleus. The presence of the dense galactic core also increases the stellar velocities near the center of the galaxy. Moreover, for small values of the distance R between the two bodies, low energy stars display chaotic motion, near the central region of the galaxy, while for larger values of the distance R, the motion in active galaxies is entirely regular for low energy stars. Their simulations suggest that in galaxies with a satellite companion, the chaotic nature of motion in not only a result of the galactic interaction between the primary galaxy and its companion, but also a result caused by the presence of the dense nucleus in the core of the primary galaxy. Theoretical arguments are presented in order to support and interpret the numerically derived outcomes. Furthermore, the authors follow the 3D evolution of the primary galaxy, when mass is transported adiabatically from the disk to the nucleus.

The authors' numerical results are in satisfactory agreement with observational data obtained from the M51-type binary stellar systems. Comparison between the present research and with similar and earlier work is also made.

In Chapter 4, a large dispersion observed in the age-metallicity relation inferred from different populations within the solar neighbourhood, is interpreted as largely due to intrinsic scatter and compared to predictions from inhomogeneous simple models of chemical evolution with stochastic star formation (Caimmi 2000). More specifically, the interstellar medium is idealized as substructured into a discrete number of $n$identical active or quiescent regions where star formation can or cannot randomly occur, respectively. The evolution is idealized as a series of identical time steps where stellar nucleosynthesis takes place within active regions, and subdivided into an assembling stage where the volume undergoes substantial changes, and a subsequent relaxed stage where the volume remains more or less unaltered. At the end of any step, a mass fraction (including newly synthesised elements) from short-lived stars is returned to the interstellar medium, instantaneously mixed with remaining gas from active and quiescent regions, and a new set of identical $n$regions is made for the next step. The theoretical age-oxygen abundance relation (TAOR) and the theoretical differential oxygen abundance distribution (TDOD) are computed for three different runs where the state of a region (active or quiescent) is decided by different sequences of random numbers. For each Galactic environment, different cases involving low ($n = 10$) and high ($n = 100$) number of regions are considered together with the case of expected evolution ($n \to +\infty$) exploited in an earlier attempt (Caimmi and Milanese 2009). Finally, the TAOR and the

TDOD are compared with the empirical age-oxygen abundance relation (EAOR) and the empirical differential oxygen abundance distribution (EDOD), respectively, inferred from subsamples made of halo, bulge, thick disk, thin disk stars and, in addition, thick + thin disk stars calculated for selected thick to thin disk mass ratios. The main results are (1) the TAOR fits to an acceptable extent (for at least one computer run) to the EAOR for both the halo, the thick disk, and the thick + thin disk, regardless of the number of regions, while the same holds for the thin disk only if the number of regions is low i.e. via stochastic effects, and (2) the TDOD fits to an acceptable extent (for at least one computer run) to the EDOD for all the above mentioned environments, regardless of the number of regions, where the fit is further improved for the thick + thin disk if the number of regions is low i.e. via stochastic effects. The results need to be confirmed using high-population, complete, and unbiased samples. An interpretation of the modelwithin current ΛCDMcosmological scenarios is shortly outlined.

As discussed in Chapter 5, the most accepted model to explain the nature of RQ/RL dichotomy of AGN is the so-called spin paradigm, which explains the jets and powerful radio emission of RQ objects as an effect of higher values of BH spin comparably to RQ ones. One of the observational differences between RQ and RL AGN predicted by spin-paradigm lies in the form of the hard X-ray primary continuum of the central machine: for the RL objects it have to be harder than for RQ ones, and have the exponential cut-off below 100 keV. However, at least several galaxies contradicting to this prescription were reported earlier, and now the authors report the similar properties of RQ Seyfert 2 NGC 4945, basing on INTEGRAL six-years observational data. NGC 4945 has an unstable spectral shape in hard X-rays, and phases with hard spectrum and exponential cut-off below 100 keV were observed here. Thus the necessity of further comparative analysis of the radio/optical/X-ray properties of such objects is stressed out.

The study of Galactic orbits in the last eighty years has been dominated by statistical assumptions made because of the lack of empirical evidence available in the early 20th century. Using evidence from Hipparcos and recent radial velocity surveys, Francis and Anderson recently showed that spiral structure is primarily a consequence of gravitational alignments of stellar orbits. Chapter 6 reviews the mechanism which creates spiral structure, considers the validity of widely held assumptions in galactic dynamics and the implications to notions such as the asymmetric drift and disc heating. It identifies a number of fundamental mathematical and physical errors in Lindblad's epicycle theory and in density wave theory. Students should be made aware that these ideas can no longer be considered as science, and authors of textbooks should consider whether they merit anything more than a historical note.

In last years, there has been important progress in the understanding of the chemicalevolution of galaxies because of the improvement of observational techniques and the development of more sophisticated numerical codes. In particular, in the Local Universe there is a tight correlation between the stellar mass and metallicity of galactic systems in such a way that brighter and more massive galaxies tend to be more metal-enriched (e.g. Tremonti et al. 2004; Lee et al. 2006). Moreover, this relation seems to evolve with time in the sense that, at a given stellar mass, systems were less enriched in the past (e.g. Erb et al. 2006; Maiolino et al. 2008).

The chemical composition of a galaxy is of fundamental importance to study its formation and evolution as it is the result of the joint action of several physical mechanisms such as the star formation processes, gas infall, Supernovae winds, the initial mass function, stellar nucleosynthesis, among others. Therefore, the determination of correlations between chemical abundances and other properties of galaxies and the study of how these relations evolve with time could provide important clues about possible galaxy formation scenarios (e.g. De Rossi et al. 2007, 2009; Finlator & Davé 2008).

In Chapter 7, the authors will review some of the most important advances made in this sense in the last decades. I will focus mainly on the study of the origin and evolution of the mass-metallicity relation as a relic from the star formation history of galaxies.

As discussed in Chapter 8, the standard model of gravitational structure formation is based on the Jeans 1902 acoustic theory, neglecting nonlinear instabilities controlled by viscosity, turbulence and diffusion. A linear instability length scale equal to the sound speed times the gravitational freefall time emerges. Because the Jeans scale LJ for the hot primordial plasma is always slightly larger than the scale of causal connection ct, where c is the speed of light and t is the time after the big bang, it has been assumed that no plasma structures could form without guidance from a cold (so LJ CDM is small) collisionless cold-dark-matter CDM fluid to give condensations that gravitationally collect the plasma. Galaxies by this CDM model are therefore produced by gradual hierarchical-clustering of CDM halos to galaxy mass over billions of years, contrary to observations showing that well formed galaxies existed much earlier. No observations exist of CDM halos. However, Gravitational instability is non-linear and absolute, controlled by viscous and turbulent forces or by diffusivity at Schwarz length scales smaller than ct. Because the universe during the plasma epoch is rapidly expanding, the first structures formed were at density minima by fragmentation when the viscous-gravitional scale LSV first matched ct at 30,000 years to produce protosupercluster voids and protosuperclusters. Weak turbulence produced at expanding void boundaries guide the morphology of smaller fragments down to protogalaxy size just before transition to gas at 300,000 years. The size of the protogalaxies reflect the plasma Kolmogorov scale with a linear and spiral morphology predicted by the Nomura direct numerical simulations and confirmed by deep field space telescope observations. On transition to gas the kinematic viscosity decreases by ten trillion so the protogalaxies fragment at Jeans scale clouds, each with a trillion earth-mass planets, as predicted by Gibson 1996 and observed by Schild 1996. The planets promptly form stars near the cores of the protogalaxies, but as the universe cools the planets gradually freeze to form the galaxy dark matter. Space telescope observations of the most distant galaxies confirm the Kolmogorov scale size and Nomura geometry of the protogalaxies. High resolution images of planetary nebula and supernova remnants reveal thousands of frozen hydrogen-helium dark matter planets with large atmospheres evaporated by such events. Galaxy mergers show frictional trails of young globular clusters formed in place, proving that dark matter halos of galaxies consist of dark matter planets in metastable clumps. Galaxy evolution is guided by friction from this galaxy dark matter.

In: Galaxies: Evolution, Morphology and Dynamics ISBN: 978-1-62081-185-6
Editors: A. Guerrera Sanchez et al., pp. 1-46

Chapter 1

ON THE NATURE OF COSMOLOGICAL ALIGNMENT EFFECTS OF GALAXIES, GROUPS AND CLUSTERS. I. OBSERVATIONS

Alejandro González-Sánchez[1], *César A. Caretta*[2], *A. Gutiérrez-Rodríguez*[1] *and P. Rivera*[1]
[1] Unidad Académica de Física, Universidad Autónoma de Zacatecas
Zacatecas, México.
[2] Departamento de Astronomía, Universidad de Guanajuato
Guanajuato, México.

Abstract

While the existence of intrinsic galaxy alignments have been repeatedly detected in a number of surveys, evidence results from two novel findings have been added that come primarily from several weak lensing studies which aims to quantify the gravitationally-induced shear field that distorts the background galaxy images. In addition, this allows for X-ray searches on substructure and the study of the dynamical state of galaxy clusters and their surroundings. This review provides an outline for the observational and theoretical evidences of the formation of coherent large-scale cosmic structures; *i.e.*, structures showing the presence of either a non-random distribution of spin vectors of neighboring galaxies, or a correlation in their morphology, or simply non-random orientations of major axes of neighboring galaxies and clusters within filaments and superclusters.

Keywords: Alignments, Galaxies, Clusters, Cosmology

1. Introduction

The diversity of alignment effects of cosmic structures detected within a wide range of scales not only has a long and rich history, dating back in time to the 1800s (Djorgovski 1987, and references therein), but also remains a controversial topic in cosmology since the first models based on the existence of dark matter were proposed. As envisioned by Djorgovski (1987) more than 20 years ago, the new recent substantial advances in cosmology –through the acquisition of more complete and reliable observational data such as the

Sloan Digital Sky Survey Data Release 5 (SDSS, DR5), the 2 Degree Field Galaxy Redshift Survey (2dFGRS) – have renewed the theoretical interest to explain the detected long-range correlations of galaxy ellipticities and orientations inferred from weak lensing studies (*e.g.*, Lee & Pen 2000a, 2000b; Heavens, Refregier, & Heymans 2000; Catelan, Kamionkowski, & Blandford 2001; Catelan & Porciani 2001; Hui & Zhang 2002). On larger scales, when studies of the dynamical state of clusters have detected substructures, these are aligned with the major axis of the host cluster and point towards the direction of large-scale filamentary structures (West, Jones, & Forman, 1995; Kolokotronis et al., 2001; Kim, 2001; Chambers, Melott, & Miller, 2000; Plionis & Basilakos, 2002, Plionis et al., 2003). Alignment effects have also been studied resorting to large N-body numerical simulations in a number of scenarios of structure formation (van Haarlem & van de Weygaert 1993; Thormen 1997; Onoura & Thomas 2000; Falthebacher et al. 2002). Observational studies, which vary in the size of the samples and methods of analysis, have produced different statistical confidence levels. Nevertheless, quite recently several groups have detected a ellipticity correlation from weak gravitational lensing studies that further reinforce the possibility of a real, non-negligible intrinsic galaxy alignment encoded in a quantifiable shape-shape correlation function (Heymans & Heavens 2002; Lee & Pen 2000a; Pen, Lee, & Seljak 2000; Brown et al. 2002; Bacon, Refregier, & Ellis 2000; Hoekstra, Yee, & Gladders 2001, 2002; van Waerbeke et al., 2000; Rhodes, Refregier, & Groth 2001). The relevance of the character of “intrinsic" for galaxy alignments, lies in the large difference in redshift space of the weak lensed galaxies, and consequently no correlation should be expected; a reasonable assumption to make. The orientation correlation of galaxy images induced by weak lensing can be quantified, but the trouble is that such a correlation cannot be reproduced if the random orientation assumption does not hold. So, a better understanding of alignment effects is needed to determine whether they can be a significant contaminant in blank fields of the statistical search for gravitational lensing, since this has been established as a powerful method to measure directly the mass distribution of the universe or the power spectrum (Bernardeau, van Waerbeke, & Mellier 1997; Hu & Tegmark 1999; Bartelmann & Schneider 1999; Munshi & Jain 2000; Cooray & Hu 2000).

In this review, we are particularly interested in the issue of a non-random distribution of the orientation of the major axes of galaxies, groups and clusters of galaxies, as a manifestation of filamentariness of the universe. The most persuasive evidence that these two aspects are closely related is that alignment effects have mainly been detected in highly flattened, large-scale structures (*e.g.*, Pisces-Perseus, Perseus-Pegasus, Virgo, L-type Abell clusters, A1656) although other coherent patterns have been reported on small-scales. Thus, different observational surveys have brought to light the existence of a wide gamut of kinds and degrees of non-random orientations, which can be classified as those present on small scales:

- between neighboring galaxies;
- between a member galaxy and its radial vector to the cluster center;
- between galaxies and their host cluster's major axis;
- between first ranked galaxies and their cluster's major axis and/or larger scale structures;

and those present on large scales:

- between galaxies within the plane of superclusters (such as the Local Supercluster and Perseus-Pisces) ranging up to $15h^{-1}$ Mpc[1].
- alignment between groups and clusters of galaxies residing in superclusters, extending up to $40 - 60h^{-1}$ Mpc (the so-called "Binggeli's effect".

Further evidence of an anisotropic orientational distribution of galaxies is the cluster-galaxy correlation function, which will be discussed later on.

2. Observational Evidences for Preferred Orientations

For each distinct alignment effect introduced here, there are different methodologies for detecting any signal of anisotropy and also many possible statistical techniques for probing the significance of the signal if it exists. Let us start by defining some usual measures of orientation of galaxies and relative angles between them or their respect to larger scale structures. Hereafter, we will refer to the orientation of the major axes of galaxies, relative to some reference direction (by convention the meridian that crosses the center of the object, angles measured from North to East), as the "position angle" (PA). The angle between the galaxies' major axes and the radius vector to the cluster center will be termed the "radial PA," and for between the major axis of a reference galaxy/cluster (such as, for instance, the first ranked galaxy) and the great circle passing through it and some neighboring galaxies/clusters the "azimuthal PA." The difference between the PAs of two objects will be called "parallel PA." Many works have considered directly the distribution of PAs (histogram) for a certain sample in order to detect parallel alignments.

Respect to the statistical methods, the most common are the parametric χ^2, Binomial and t-Tests, the non-parametric Kolmogorov-Smirnov and Kuiper Tests, the Wilcoxon rank-sum Test, the Rayleigh Test (all for estimating the probability that the observed distribution could be originated from a random isotropic parent distribution), the Fourier Transform of the distribution, the First Autocorrelation Coefficient and the Cross Correlation, as well as comparing the results with a set Monte Carlo random realizations (see, e.g. Hawley & Peebles 1985; Djorgovski 1983; Fong et al. 1990; Cabanela & Aldering 1998; Chambers et al. 2001; Faltenbacher et al. 2007 for descriptions of the tests). Fortunately, different tests usually agree well on estimation for the significance in the results for this kind of data.

Next, we will sketch and discuss the main observational and theoretical studies that have shown reliable evidence in favor of alignments of galaxies in clusters.

3. Small-Scale Alignment Effects

We will consider in this section the observational evidences for alignment effects on scales smaller than the size for typical clusters of galaxies. Most of the studies have focused

[1]Note that we adopt in this paper the standard cosmological notation of writing the Hubble parameter, $H_0 = 100h$ km s^{-1} Mpc^{-1}.

the environment of rich clusters of galaxies, although some results for galaxy groups and isolated spiral galaxies have also been obtained.

3.1. Radial and Parallel Anisotropies Inside Galaxy Clusters

The first systematic studies on the existence of preferred orientations for the major axis of the galaxies date from the dawn of Extragalactic Astronomy (Brown 1938, 1939; Wyatt & Brown, 1955; Reaves, 1958; Brown, 1964; Kristian, 1967; Rood & Baum, 1967; Gainullina & Roshjakova, 1967; Brown, 1968; Reinhardt, 1971, 1972; Rood & Sastry, 1972; see also Hawley & Peebles, 1975 and Djorgovski, 1987 for reviews of the earliest period). At that time the available data came from photographic plates (such as the ones from the Palomar Observatory Sky Survey, POSS), while shapes and orientations were measured by eye, using overlays and protractors. Some signs of preferred orientations were found in the regions of Horologium, Cetus, Pisces, Hydra, Virgo, Coma, and Corona Borealis, all containing relatively nearby galaxy clusters.

Two carefully controlled observational studies on alignment effects were first performed by Hawley & Peebles (1975) and Thompson (1976). The former authors found an anisotropic distribution in the galaxy orientations in the Ursa Major and the Virgo clusters but it was just above the statistical uncertainty. In the case of Coma, there was no evidence of anisotropy relative to the cluster major axis. However, the galaxies showed a trend for radial alignment with the cluster center. >From eight rich Abell clusters: Virgo, A119, A400, A1656 (Coma), A2147, A2151 (Hercules), A2197 and A2199, Thomson (1976) confirmed the tendency for radial alignment of the elliptical galaxies in the Coma cluster (see also Thompson & Gregory, 1978; Gregory & Thompson, 1978), and a significant trend for parallel alignments in A2197 – being more pronounced for galaxies detected in red photographic plates. These findings gave support to previous studies, with less accurate methods of analysis, by Gainullina & Roshjakova, (1967), who detected radial alignments for A1656 and A2065 (Corona Borealis) and Rood & Sastry (1972), who found a tendency for parallel alignment in A2199.

The advent of the automatic plate measuring machines made possible to extend the analysis to large samples of galaxies, probable members of the galaxy clusters, and also for studying fairly ample samples of clusters. For instance, Djorgovski (1983a,b; 1987), in a subsequent study of Coma confirmed the tendency for radial alignment of galaxies. Moreover, his analysis of 4,217 galaxies allowed more detailed conclusions: red galaxies show a more prominent alignment than blue galaxies, fainter galaxies are more likely to be aligned than the brighter ones, and the orientations of the galaxies in the cluster core (closer than about 0.8 h^{-1} Mpc) are less anisotropic than the ones for galaxies at larger distances. A considerable parallel alignment was also found for three in a set of eight linear shaped clusters (type L, Rood & Sastry, 1971, hereafter RS): A426, A999 and A2197, by Strom & Strom (1978) and Adams et al., (1980). MacGillivray & Dood (1979a,b) also found some signs of preferred orientations in the southern clusters S141 (tangential, i.e., anti-radial) and S84 (radial). Tangential alignments were also found before by Rood & Sastry (1971) for the clusters A154 and A1185. Modestly significant results for preferred orientations, both radial and parallel, were found by Fong et al., (1990), for a sample of 621 galaxy clusters.

However, some negative results were also found. Reaves (1958), Kristian (1967), Rood

& Baum (1967), Kindl (1987), van Kampen & Rhee 1990, and Trevese et al., (1992) all failed in detecting anisotropies in the orientations of cluster galaxies (except for the brightest ones, as shown later).

New technologies and techniques came up in the last two decades: Charge-Coupled Devices (CCDs), extensive redshift surveys (such as the 2dF Galaxy Redshift Survey - 2dFGRS, e.g., Colless et al., 2001, and the Sloan Digital Sky Survey - SDSS, e.g., Abazajian et al., 2009), catalogs of clusters detected in X-rays, and also direct measurements of galaxy distances, which have improved considerably the studies of cosmic alignments. Although Torlina, De Propris, & West (2007) found no significant deviations from isotropy in the orientations of 188 galaxies (60spectroscopically confirmed members) of Coma cluster, many recent studies were able to give some light on the strength and nature of the alignment effects. Apart from the convincing alignments found for the brightest elliptical galaxies in Coma (West, 1998) and Virgo (West & Blaskeslee, 2000), and the statistically significant radial alignment observed in 85 X-ray selected clusters (Pereira & Kuhn, 2005), Plionis et al., (2003) and Faltenbacher et al., (2007) present robust results from large samples of clusters and groups of galaxies. Faltenbacher et al., (2007) studied 18,576 groups and clusters of galaxies, detected in SDSS-DR4, and found that: (i) the largest signal of anisotropy is presented by the first-ranked galaxy, which is usually aligned with the orientation of the surrounding distribution of galaxies (see the later discussion); (ii) for the region inside a radius of 0.7 times the virial radius of the system, the red galaxies have a significant tendency to point toward the first-ranked galaxy (radial alignment); (iii) for distances smaller than 0.1 times the virial radius the galaxies present a weak tendency to be aligned in parallel with the first-ranked galaxy. Plionis et al. (2003), studying 303 clusters, also found that the strongest anisotropy is presented by the first-ranked galaxy, followed by the 10 next brightest members. More interesting, they found that galaxy members alignment appears to be stronger the more dynamically young the cluster is, especially if it presents high velocity dispersion and is located in a high density environment (such as that of the superclusters of galaxies). They corroborate their result with a deeper analysis of the cluster A521. Also, in the same direction, Kitzbichler & Saurer (2003) found parallel alignments related to substructures in Coma cluster.

Nevertheless, what could be the cause of the conflicting results and low significance of the signal, when detected, for the radial and parallel alignment of galaxies in clusters/groups? Firstly, one should consider that most systematic biases, such as the projection effects, always work in the direction of smearing out any true alignment signal. False anisotropies may also be generated by such distortions (for instance the effect of bad telescope tracking) but these are usually identifiable and can be modeled and eliminated. Thus, the significance level of some detected anisotropy signals may be considered as a lower limit. Works prior to the advent of the automatic scanning of photographic plates may have suffered from systematic observer biases (also called physiological biases), as argued by Öpik (1968), since the measurements were done by eye, resulting in a critical dependence on researcher's judgment. The measuring machines are also prone to systematic bias, as can be clearly seen in the data used by Djorgovski (1983), presented in his Figure 1 (see also Figure 1 in Kitzbichler & Saurer 2003). Another problem that may also probably weaken the alignment signal is the mixing of different galaxy morphological types. There is some evidence of a real morphology-alignment effect, which will be discussed later, that produces different alignment patterns for ellipticals and disk galaxies, for example. This may be the cause of the isotropic results or small anisotropic ones found by Rood & Baum (1967), Hawley & Peebles (1975), Kindl (1987), Fong et al. (1990) and Torlina et al. 2007. Furthermore, there is possibly an effect of weaker (or confusing signal of) anisotropy in the cluster core, as pointed by Djorgovski (1983). Hao et al., (2011) also found an inconsistent signal in the low redshift bin of their data (associated with larger nearby systems), and proposed two possible explanations: contamination by the diffuse light from the brightest cluster galaxy, affecting the PA measurement of its closest neighbors, or some tidal torque twisting effect in the same central galaxy. Finally, as already discussed, the dynamical state of the cluster/group may strongly affect the orientations of the member galaxies. Taking into account a large radius may include different substructures and filaments, each one possibly with a particular preferred orientation. Concerning the dynamics in the clusters of galaxies, González & Teodoro (2010), with a simple model opened the possibility that dynamical friction of massive galaxies could contribute to the generation of radial and parallel alignments in central and regions of the host cluster.

3.2. Alignment of Neighboring Galaxies

Reciprocal tidal torques (e.g., Kitzbichler & Saurer 2003) may induce alignments between close pairs of galaxies. This possibility was first systematically probed by Hawley & Peebles (1975), using a catalog of 5,559 galaxies, selected from photographic plates, in the region of three clusters of galaxies (Ursa Major, Virgo, and Coma). They found no evidence of anisotropy in the relative orientations of galaxies separated by less than about 1 h^{-1} Mpc in projection. A different result was found for Coma cluster (A1656) galaxies by Djorgovski (1983), who found that galaxies tend to point to their nearest projected neighbors. More recently, A1656 was revisited by Kitzbichter & Saurer (2003) and Torlina et al. (2007). While Torlina et al., considering mostly spectroscopic members and using shape parameters obtained from CCD photometry, found no significant deviation from isotropy, Kitzbichter & Saurer detected neighbor alignments on small scales, probably associated with substructures identified from the subtraction of a β-model to an XMM-Newton hot intracluster gas image.

The main problem affecting these ambiguous results is probably the limitation imposed by the use of projected information: both PAs and vicinity cannot be uniquely defined, which may confuse any alignment signal. Also, a morphology dependent alignment effect may be taken into account, as will be discussed further. Finally, such studies were restricted to almost only one cluster, and other systems and different environments must be investigated in order to clarify this question and effectively measure and characterize this possible close pairs alignment.

Table 1. Observational searches for the alignment of the CDGs.

Work	CDGs		Host Cluster/Group		$N_{<30}/N$ (%)[d]
	data source[a]	selection[b]	N systems	selection[c]	
Sastry (1968)	POSS (manual)	cD	9	cD-type	5/5(100)
Dressler (1978)	Lick plates/scanning	cD	12	cD-type	4/5(80)
Carter & Metcalfe (1980)	various plates/scanning	cD	22	cD-type	13/14(93)
Binggeli (1982)	POSS (manual)	BCG	44+12	BM≤II+BM>II	33/51 (65)
Struble & Peebles (1985)	POSS (manual)	BCG	237	in SCs	125/198(63)
Rhee & Katgert (1987)	POSS (Astroscan)	CDG	37	B82+SP85	24/34(71)
Struble (1987)	CCD (various)	BCG	21	w/ isoph. photom.	12/17(71)
Tucker & Peterson (1988)	CCD	BCG	27	+MKW/AWM	13/26(50)
Rhee & Roos (1989)	POSS (Astroscan)	BCG	108	z<0.1	24/43(56) [all]
					18/29(62) [cD+E/S0]
Struble (1990)	POSS (manual)	2BCGs	68	B-type	35/66(53) [1C]
					50/68(74) [RC]
van Kampen & Rhee (1990)	POSS (Astroscan)	BCG	122	R≥1, z<0.1	65/112(58) [all]
					22/30(73) [cD]
Porter et al. (1991)	CCD	BCE	175		17/22(77) [$38h^{-1}$ isoph.]
					21/40(53) [$5h^{-1}$ isoph.]
Trevese et al. (1992)	POSS (PDS)	BCG	55	+Zw	30/50(60) [1C]
					27/50(54) [RC]
Fuller et al. (1999)	CCD	BCG	21	MKW/AWM	8/13(62)
Kim et al. (2002)	SDSS-EDR	BCG	115	MF+MaxBCG	39/66(59) [$\Delta m \geq 0.65$]
					17/49(35) [$\Delta m < 0.65$]
Stazzullo et al. (2005)	DPOSS	CDG	28	Compact Abell	14/22(64)
Yang et al. (2006)	SDSS-DR2	CDG	24,728	NYU-VAGC	
Han et al. (2011)	SDSS-DR7 (z_{phot})	BCG	~11,000	GMBCG	

a) Acronyms: POSS (Palomar Observatory Sky Survey); CCD (Charge-Coupled Device); SDSS (Sloan Digital Sky Survey); DPOSS (Digitized POSS).
Other codes: "manual" (by using overlays, rulers, protractors and eyepieces); "Astroscan" (Leiden Astroscan plate measuring machine); "PDS" (PDS 1010G microdensitometer); "z_{phot} (made use of photometric redshifts).
b) see the text for the acronyms of CDGs types.
c) "-type" [Rood & Sastry (1971) rich clusters classification scheme]; "BM" [Bautz & Morgan (1970) rich clusters classification system]; "SCs" [superclusters of galaxies]; "B82+SP85" [clusters both in Binggeli (1982) and Struble & Peebles (1985) samples]; "MKW/AWM" [poor clusters from Morgan et al. (1975) and Albert et al. (1977)]; "Zw" [Zwicky et al. (1968) clusters]; "MF+MaxBCG" [clusters detected by both Matched Filter (e.g. Kim et al. 2001) and MaxBCG (Annis et al. 2001) techniques]; "NYU-VAGC" [New York University - Value-Added Galaxy Catalog groups (Blanton et al. 2005)]; "GMBCG" [Gaussian Mixture Brightest Cluster Galaxy Catalog (Hao et al. 2010)].
d) "1C" (alignment between first-ranked galaxy and its parent cluster); "RC" (alignment between the radius-vector connecting the two brightest galaxies and their parent cluster); "Δm" (magnitude gap between the BCG and the average of the second and third brightest cluster galaxies).

3.3. Alignment between First-Ranked Galaxies and Their Parent Groups/Clusters

The most firmly established alignment tendency is the one presented by the Central Dominating Galaxies (CDGs), which are usually the first-ranked galaxies in clusters or groups and present an early-type morphology (cD, D, gE, or simply the brightest E). This distinction is necessary in order to separate this anisotropy from the Holmberg's effect that will be discussed next. A representative sample of the studies of the preferred orientation of CDGs is shown in Table 1 (see also Figure 1).

The anisotropy signal is commonly measured by the acute angle between the PA of the CDG and the major axis of the cluster/group galaxy distribution or the major axis of its X-ray image (azimuthal PA). Both optical (from galaxy distribution) and X-ray PAs are subject to larger inaccuracies than the PA of the CDG. Uncertain and sub-sampled membership and choice of the initial center and shape (usually certain circular area and a redshift interval) are the main problems for defining optical orientation. For X-ray emitting gas, on the other hand, the main limitations come from the low signal-to-noise ratio for faint clusters, restricted instrument FOV for nearby clusters, and the choice of the profile fitting model. Both suffer from the difficulty in separating substructures and excluding superposing systems along the line o sight. The errors in both are usually estimated to be in the range 15–30°, while the galaxy PAs usually have uncertainties around 5°.

All the works listed in Table 1 find significant signals of alignment, confirmed by different statistical tests. In order to compare their results we show, in column 6, the fraction of CDGs that present an azimuthal PA of less than 30° relative to the total number of studied systems. Remarkably, the cD galaxies present the largest anisotropy signals: the fraction of aligned cDs, according to the results presented on the table, is around 70-90%.

By putting together all the BCGs, irrespective of their dominance, the fractions are around 50-70%. Struble & Peebles (1985), for instance, found an alignment of 88for the RS type cD, around 70% for types B and C, and around 50% for types L, F, and I. This dependence of the alignment of the CDG on its dominance was also found in the results by Rhee & Roos (1989), van Kampen & Rhee (1990), Kim et al. (2002), Trejo-Alonso (2008) and Niederste-Ostholt et al. (2010). Some works have also pointed out that in B-type clusters, besides the alignment of the brightest member of the dominant pair, the line that connects the two CDGs is often aligned with the major axis of their host cluster (again between 50 and 70% of the time, see Struble, 1990; Trevese et al., 1992; and also West, 1998 and West & Blakeslee, 2000). As an example Binggeli, Tammann & Sandage (1987) have noted that "*the line connecting M87 and M84 appears a fundamental axis of the cluster*." Moreover, some studies have shown that the second brightest galaxy is also propense to be aligned with its parent cluster, although with a weaker signal (Struble 1990; Trevese et al. 1992; Torlina et al. 2007). Another important consideration is the surface brightness level of the CDG's elliptical isophote taken into account. The figure 5 of Tucker & Peterson (1988) illustrates this dependence (see also Porter et al. 1991): internally the CDGs tend to be rounder, while at larger radii the isophotes become predominantly flatter and strongly aligned with the orientation of their host clusters/groups. This result suggests that at least the external regions of the CDGs are probably dynamically coupled to their host systems.

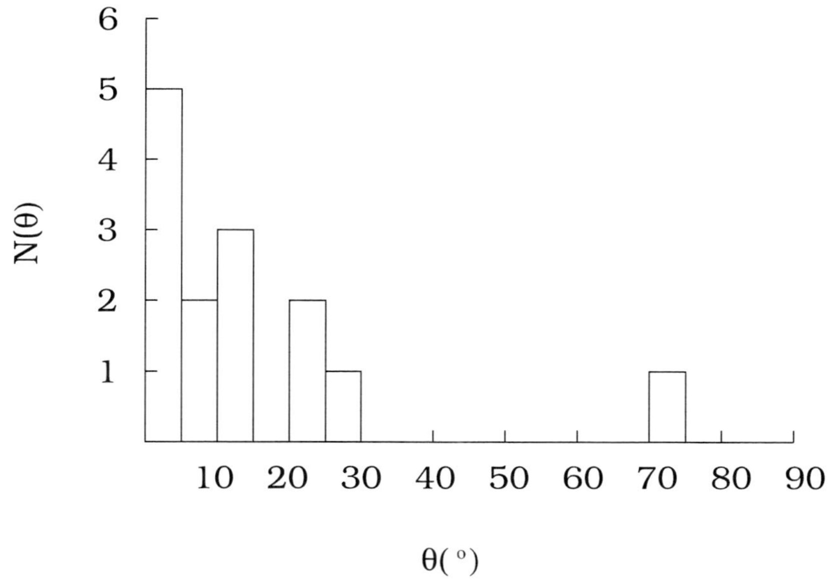

Figure 1. Histogram showing the orientation of cD galaxies relative to the major axes of their host cluster. From Carter & Metcalfe (1980).

3.4. The Holmberg's Effect

As defined by Holmberg (1969) this is the study of the distribution of satellite galaxies (usually dwarfs) around a relatively isolated giant spiral galaxy. The relatively isolated means that this includes giant spiral galaxies that are members of small groups, such as the Local Group.

The modern studies of this effect consider that the primaries (also called hosts) must be at least 2.5 magnitudes brighter than any other galaxy that falls within a cylinder of $R_p = 700\ h^{-1}$ kpc and $|dv| = 1{,}000$ km s^{-1}, to guarantee the isolation, while the satellites are required to be at least 4 magnitudes fainter than their primary and be located inside another small cylinder of $R_p = 250\ h^{-1}$ kpc and $|dv| = 500$ km s^{-1}. This definition clearly excludes potential primaries that are members of groups and clusters of galaxies. On the results of cosmological N-body simulations containing dark matter, the halos of isolated galaxies, groups and clusters present the same structure, although in different scales (see, for instance, Moore et al., 1999). Since people are used to referring to all of them as halos, and to their sub-haloes as central object and satellites, some confusion has occurred in the literature. Obviously, in terms of baryonic matter objects, clusters, and groups of galaxies are very different from isolated spiral galaxies (in fact, there is a "missing satellites problem," still unsolved, that prevents the correct concordance between observations and simulations). An example is the study by Yang et al., 2006, which advocates to discuss the Holmberg's effect but, in reality, their observational sample is composed by groups and clusters of galaxies.

What Holmberg found was a tendency of the satellites of isolated giant spiral galaxies, within a radius of $R_p = 40\ h^{-1}$ kpc, to be distributed along the elongation of the minor axis of their primaries. A striking example of the Holmberg's effect is our own Milky Way

Galaxy (e.g., Majewski 1994; Kroupa, Theis & Boily 2005), which have its nearby satellites distributed in a polarly elongated manner.

Some studies have found evidence in favor of the Holmberg's effect. Zaritsky et al., (1993) found that with a small number of satellites within $R_p \leq 75\ h^{-1}$ kpc, that they seem to avoid small azimuthal PAs. However, beyond this radius, the distribution of angles tend to be uniform. Interestingly, above $R_p \leq 250\ h^{-1}$ kpc there is an apparent new external Holmberg's effect (see their Fig. 8). In a subsequent work with a larger sample, Zaritsky et al., (1997b) do not confirm the internal effect, but find a statistically significant signal for a Holmberg's effect between 200 and 400 h^{-1} kpc from the primaries. Another positive detection for the Holmberg's effect was carried out by Sales & Lambas (2004) using a sample extracted from the 2dfGRS.

Nevertheless, many other works failed in finding evidence of the effect. Voltanen et al. (1978) found an excess of compact satellites with small azimuthal PAs, and suggested that Holmberg's result was due to not considering these objects. Brainerd (2005), Agustsson & Brainerd (2006a) and Azzaro et al. (2007) also found isotropic results for isolated blue primaries from the SDSS, while, for red primaries the effect seems to approach the one presented by groups and clusters of galaxies (the alignment of the major axis of the brightest galaxy with the distribution of its "satellites").

4. Large-Scale Alignment Effects

Since member galaxies seem to present some propensity to be aligned with their parent groups/clusters, the question arises if they can be also aligned with the large scale structures in which their host systems are embedded.

There is increasing evidence that alignments occur on large scales in the Universe. Although many cosmologists would say that cluster-cluster alignments could be present on scales of up to $60h^{-1}$ Mpc, the evidence is much clearer and better supported up to $30 - 40h^{-1}$ Mpc. Some observational studies on the scale of clusters of galaxies, involving cD galaxies and substructure for example, suggest that there is a correlation between them and the large-scale density field. In this section, we will review the observational grounds for alignments on such scales.

4.1. Anisotropic Orientations on Scales of Superclusters and Filaments

Three superclusters of galaxies have been especially scrutinized for searching anisotropies in the orientation of their member galaxies: the Local Supercluster (Reinhardt & Roberts, 1972; Helou & Salpeter, 1972; Jaaniste & Saar, 1977; De Vaucouleurs, 1978; MacGillivray et al., 1982; Kaprandis & Sullivan, 1983; Dekel, 1985; MacGillivray & Dood, 1985; Flin & Godlowki, 1986; Flin, 1988a; Kashikawa & Okamura, 1992; Han et al., 1995; Flin, 1996; Hu et al., 1998; Godlowski & Ostrowski, 1999), The Perseus-Pisces Supercluster (Brown, 1964; Gregory et al., 1981; Vettolani & Pillastrini, 1987; Flin, 1988b; Cabanela & Aldering, 1998) and Coma/A1367 supercluster (Flin, 1993; Wu et al., 1997; Flin et al., 2001).

The results are inconclusive since some authors found distributions of PAs consistent with isotropy (De Vaucouleurs, 1978; Dekel, 1985; Vettolani & Pillastrini, 1987; Han et al., 1995; Godlowski & Ostrowski, 1999) while others found bimodal distributions of PAs or marginally significant signals of alignment. Reinhardt & Roberts (1972), for example, found that the major axes of spiral galaxies tend to avoid the direction of the LSC poles, and detected a tendency for galaxies at low supergalactic latitudes to have larger apparent ellipticities. This effect is expected if galaxies really are aligned along the LSC plane. Kapranidis & Sullivan (1983) have also confirmed that there is a non-random effect present, in which galaxies are aligned preferentially along the the LSC. Further evidence of alignment of galaxies in superclusters was given by Gregory et al., (1981) who analyzed the orientation of galaxies in the Perseus (A426), A347, A262, NGC383, and NGC 507 clusters, which are joined in the Perseus supercluster. This supercluster has a filamentary component whose depth is approximately the same size as its minor axis width. A statistically significant bimodal distribution (at the 1.6% confidence level) was found for the galaxy PA; one peak coincided with the PA of the supercluster filament.

The complexity of the situation increases when samples considering both PA and ellipticity information, in order to determine the spin vector direction, are considered. Flin & Godlowski (1986), for instance, who analyzed the orientation of 1,275 galaxies in regions close to the plane of the Local Supercluster, found an anisotropic distribution of orientations in which the the planes of galaxies tend to be perpendicular to the LSC plane; the galaxy rotation axes are aligned with the supercluster plane. The preferred orientation of the galaxies was strongly dependent on the sample of galaxies, *i.e.*, the orientation of face-on galaxies is different to that of edge-on galaxies. As the major number of these galaxies are oriented close the LSC plane, the rotation axes of face-on galaxies tend to be aligned with the LSC plane. Also, if these galaxies are extracted from the sample the edge-on galaxies yield an excess of galaxy rotation axes perpendicular to the LSC. They pointed out that this dependence of the orientation on morphological type was a curious phenomenon not easy to explain. Flin (1988a), using radial velocities to determine galaxy membership of superclusters, and only disc galaxies – for which rotation axes are well determined – found the same tendency in the three superclusters mentioned before. Besides the search for alignment in a supercluster as a whole, he also looked for the effect separately in the main structure of the supercluster – namely the disc of the LSC, clusters A1656 and A1367 in the Coma supercluster and the main ridge of the Perseus supercluster – and the remaining part of a supercluster. He found that within each supercluster, galaxy planes tend to be perpendicular to the main plane of the supercluster. Also, the alignment is observed both in the dispersed component of superclusters and in densely populated regions. For the same supercluster, particular substructures can have different orientations, and the projections of galaxy rotation axes tend to point towards the main structures. Most other investigations with these kind of data (Jaaniste & Saar 1977; Flin 1988b, 1993, 1996, 2001; Hu et al. 1998) also found that the spin vectors of lenticulars and spirals tend to be aligned along the plane of the LSC.

Furthermore, different alignment signal for galaxies near to the LSC plane and galaxies high above it were detected (Helou & Salpeter, 1982; MacGillivray et al., 1982; Kashikawa & Okamura, 1992). Specifically it appears that disc galaxies close to the plane show their spin vectors parallel to it or randomly oriented, while disc galaxies in the outskirts of the plane have perpendicular spin vectors.

Another way to verify if the galaxies "feel" the distribution of matter in larger scales is by correlating the orientation of galaxies directly with the cosmic filamentary pattern.

Seldner & Peebles (1977), Argyres et al., (1986) and Lambas et al., (1988b) noted that the count of galaxies in the Lick catalog are systematically higher in the direction of the major axes of the brightest galaxies in rich clusters. This kind of alignment was found to extend up to at least $15h^{-1}$ Mpc, and there exists weaker evidence that the alignment extends up to $30h^{-1}$ Mpc.

A posterior study by Muriel & Lambas (1992) searched for systematic effects in the orientation of galaxies with respect to their neighbors on different scales. They used different samples of the ESO catalog of galaxies. In samples of spiral galaxies, statistical evidence (2.8σ) was found for a preferred orientation of the major axis in the direction toward the nearest neighbors on scales of $3h^{-1}$ Mpc. Elliptical galaxies were found to be preferentially oriented not only with the nearest but also with all neighbors on scales of $3h^{-1}$ Mpc. Anisotropies in the orientation of spiral galaxies in samples of the Uppsala, ESO/Uppsala and FGC (Flat Galaxies Catalog, Karacehntsev et al., 1993a) catalogs have been also reported by Mandzhos, Tel'nyuk & Gregul' (1985) and Parnoviski et al., 1994).

4.2. The Binggeli's Effect

In the pioneering analysis by Binggeli (1982), optical positions of the 50 brightest galaxies within 1 h^{-1} Mpc of the centers of 44 rich clusters were used to measure the ellipticity of their galaxy distribution and to define the respective major axes. Binggeli (1982) found that the major axes of clusters whose nearest neighbors were closer than $\sim 15h^{-1}$ Mpc subtended in every case an azimuthal angle of less than 45°. While the alignment of neighboring clusters would favor the tidal mechanism, Binggeli (1982) found that, by comparing the orientation of clusters not only with their nearest neighbors but also with each neighbor in a surrounding volume of radius $50h^{-1}$ Mpc, an alignment effect on a very large-scales emerges (Figure 7), suggesting two things: tidal distortions cannot be the main process to cause the preferential orientation of clusters, and that superclusters are real physical structures.

Soon after, Struble & Peebles (1985) put in doubt Binggeli's results when, by using an extended sample composed by clusters that were probable members of superclusters, found just a marginally statistically significant sign of alignment, at a level that might be influenced by systematic errors. Other studies also failed to confirm the effect (Ulmer et al. 1989; Martin et al. 1995), or found that the signal was not as convincing as suggested by Binggeli's results (Rhee & Katgert 1987; Fong et al. 1990; Lombard & Cohen 2006).

However, most of the works in the field give support to the tendency that neighboring groups and clusters to 'point to each other' (Fontanelli 1984; Flin 1987; Rhee & Roos 1989; West 1989a,b; Lambas et al. 1990; Rhee et al. 1992; Plionis 1994; West 1994, 1995; Fuller et al. 1999; Chambers et al. 2000, 2002; Plionis & Basilakos 2002; Wang et al. 2009).

The large-scale alignments discovered by Binggeli (1982), were found by comparing the orientation of the clusters with the distribution of their surrounding clusters within a spherical volume. Accordingly, one can expect the alignment to show up in an even more pronounced way if the orientations of the clusters were exclusively compared with those of their host superclusters.

Positive results were found in this manner for clusters in the Perseus supercluster by

Einasto et al. (1980), and for clusters in various large-scale filamentary structures, defined by the distribution and orientation of double galaxies by Tifft (1980). West (1989b), Rhee et al. (1992) and Plionis (1994) have generated supercluster catalogs from samples of hundreds of galaxy clusters in the nearby Universe ($z \leq 0.1$) in order to study the Binggeli's effect inside superclusters. The superclusters were identified using a percolation algorithm (also known as Friends-of-Friends, or FoF, technique), and the results revealed that about 70% of the clusters they analyzed probably reside in superclusters. By comparing the alignment signal for clusters within superclusters with that for nearest neighbors irrespective of their possible supercluster membership, coincide that, although both samples present significant alignment signal, the forming sample show the strongest one.

What has been discussed was the extent of the effect, and ranges between 10 and 30 h^{-1} Mpc have been found.

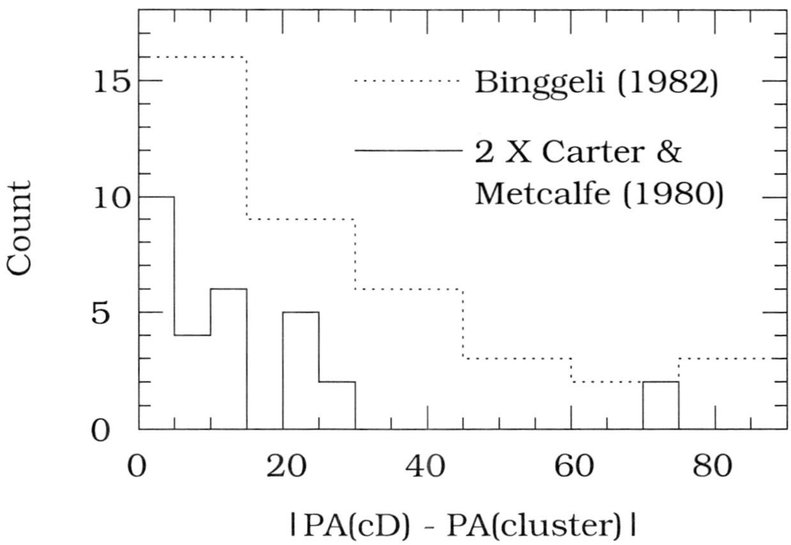

Figure 2. The PA difference between the cluster PA and the cD galaxy, showing the strong alignment effect. From Djorgovski (1987).

4.3. Synthesis of the Probable Correlations

Many observational studies have pointed that the alignment signal may depend on the different morphological types of the galaxies (Thompson 1976; Strom & Strom 1978; MacGillivray & Dood 1979b; Djorgovski 1983a,b, 1986; Lambas 1988a; Muriel & Lambas 1989; Rhee & Roos 1989; West 1998; Cabanela & Aldering 1998; West & Blakeslee 2000; Plionis et al., 2003; see also the papers about the orientation of the spin vectors of spirals in section 2.2.1).

The first strong evidence of this effect came from the work of Lambas, Groth & Peebles (1988a), who have studied the orientation of galaxies of different morphological types relative to the large-scale structures in which the galaxies are embedded. They used a sample from the Uppsala General Catalogue (UGC, Nilson 1973) of bright galaxies. Although the morphological type information in the UGC was ambiguous, for example galaxies may be classified elliptical E, E-S0, S0-E and so on without any distinction, galaxies were classified only as ellipticals, lenticulars and spirals according to whether the first morphological type listed was E, S0, or S respectively.

Not only an important alignment effect was detected between galaxies in larger structures extending up to $\sim 4 - 5h^{-1}$ Mpc, but the effect was also galaxy-morphology dependent. Alignments were detected for elliptical galaxies but not for spirals and lenticulars (disc galaxies). This dependence cannot be explained in terms of a very probable misclassification in the morphological type of the galaxies. Were this so, the misclassification of spirals as ellipticals would dilute the true alignment effect, but a misclassification of elliptical as spiral galaxies would not introduce a spurious alignment effect for spirals. If the tidal field were responsible for the alignment of elliptical galaxies, then in the same way, this tidal field must have been responsible for the origin of the angular momentum of the spiral and lenticular galaxies. Therefore, one should expect the same degree of alignment in disc and spheroidal galaxies, or a correlation in the orientation of neighboring galaxies, which was not observed.

We recall that in the numerical simulations of Barnes & Efstathiou (1987) on angular momentum acquisition of the perturbations through torques, it was found that tidal torques do not produce alignment of perturbations, as indicated by the lack of a spin-vector versus separation correlation between neighboring galaxies. The tidal effect from the far density field could probably be more important. Such a conclusion was formulated by Lambas et al. (1988a,b) who pointed out that it is plausible that the orientation of elliptical galaxies reflects the primordial orientation of maxima of the density field. In this case, the angular momentum of disc galaxies need not have any connection with the orientation of elliptical galaxies. Further, the density peaks themselves would be oriented not only in the same direction but also they would be oriented by the direction imposed by the large-scale tidal shear of the structure to which they belong.

It has been long known that spheroidal galaxies (such as E and D) are predominantly supported by internal random motions of stars (e.g., Bertola & Capaccioli, 1975, see also Merritt, 1999 for a review), while disc objects (such as S0 and S) are rotationally supported systems. This was not known when the first studies of preferred orientations were carried out, which led many researchers to consider a direct relation between ellipticity and rotation for all galaxies. Also, many works have put E together with S0, which potentially blurs the alignment signal if it is different for spheroidal and disc galaxies. Taking this factor into account, or not, in the morphology of the galaxies; may affect considerably the results. Since the mechanisms that produce these alignments may be very different for rotating disks and velocity dispersion supported systems, especially if they are dominantly evolutive. As will be discussed later, the conservation of intrinsic angular momentum makes the orientations of disc galaxies much more stable than the orientations of spheroids.

The effect of luminosity (see, e.g., Djorgovski 1983b), which is a proxy for the mass of the galaxies, although clear since the galaxies that present the largest alignment signals are the giant ones (such as cDs and other CDGs), is not well understood.

The third important correlation, which may be very important for distinguishing between primordial and evolutive galaxy alignments, is the local density. Primordial alignments should remain only in low-density regions, since the galaxy-galaxy interactions in dense regions would easily smear out any primordial signature. Conversely, any alignment generated by tidal torques or dynamical friction (shown later) should be stronger in high-density regions, since they require multiple galaxy-galaxy interactions or a strong tidal field (e.g., Cabanela & Aldering 1998).

Nevertheless, it seems that, more than the local density itself, the local dynamical activity is probably more critical in affecting the orientations of the galaxies and their systems. As mentioned before, Plionis et al., (2003) noticed that galaxy members alignment appears to be stronger the more dynamically young the cluster is, especially if it presents high velocity dispersion and is located in a high density environment (as that of the superclusters of galaxies), while Kitzbichler & Saurer (2003) found a direct correlation between the alignment of galaxies and their membership to substructures. In the scale of cluster-cluster alignments, Plionis & Basilakos (2002) found that by estimating the two-point correlation function for APM clusters, the clusters with more substructuring are significantly more clustered than the complete sample. It was also found that there was a strong correlation between the strength of the alignment signal and the substructure significance level (a proxy of the dynamical youth of the system). This concluded that the more aligned neighboring clusters were the ones showing evidence of dynamical activity and residing in high-density environments, such as that of superclusters.

As a consequence, the alignment effect may not be correlated directly with the mass of the group/cluster. Evidence of this are given by the fact that CDGs in poor clusters or groups show the same or possibly even more stronger signals of alignment than CDGs in rich clusters (West 1989a; Fuller et al., 1999; Yang et al., 2006). In fact the most dominant CDGs are not necessarily in the centers of the most massive clusters of galaxies, in contrast to what is usually expected from the N-body dark matter simulations. There is now much evidence in favor of the preferred evolution of dominant galaxies in groups, with their smaller velocity dispersions, than in rich clusters (see, e.g., Coziol et al., 2009 and references therein). These dominant galaxies, then, migrate to the centers of the clusters when the groups that form hierarchically the clusters are merged.

A final important correlation to be pointed is the evolution of the alignments with redshift. Very recent works have noted that the anisotropy in galaxy orientations was already present at relatively high redshifts. Donoso et al. (2006) found that the major axes of Luminous Red Galaxies (LRGs) are preferably aligned with the distribution of red tracer-galaxies up to 1.5 h^{-1} Mpc, in a sample with $0.4 < z < 0.5$. A even more encouraging result was presented by Hao et al., (2011): their research found, by using a volume limited sample of $\sim$ 11,000 clusters of the GMBCG catalog, derived from the Data Release 7 of SDSS, that the alignment signal becomes stronger as the redshift decreases (see their figures 10 and 14). This clearly points to the effect, at least partially, of evolutive mechanisms in generating the alignment of galaxies.

4.4. Correlation Function and Lensing Effect

The anisotropy in the distribution of galaxies is quantified through the so-called cross-correlation function ξ_{cg}, which measures the excess of probability, over random, of finding a galaxy within a given separation r of the cluster – *i.e.*, it describes the enhanced density of galaxies around a cluster. Bahcall (1987) fitted the correlation observed by a power law model for the function (Peebles, 1980)

$$\xi_{cg} = (r/7h^{-1}\text{Mpc})^{-2.5} + (r/12.5h^{-1}\text{Mpc})^{-1.7} \qquad \text{for } r < 40h^{-1}\text{Mpc} \ . \qquad (1)$$

The first term of this correlation function represents the internal density of galaxies in a cluster. The more slowly varying contribution of the correlation function found at large scales, and represented by the $r^{-1.7}$ part, is produced by the clustering of clusters – *i.e.*, clusters provide an excess in the concentration of galaxies near neighboring correlated clusters; this was interpreted by Bahcall (1987) as indirect evidence that protostructures possess long tails along their major axes, which extend to large scales, *i.e.*, connecting bridges between neighboring structures (Batuski & Burns, 1985). Observational work by Flint & Impey (1996) reinforces this possibility since they detected an anisotropic distribution of galaxies behind the Virgo cluster, mainly concentrated in three main galaxy peaks lined up and pointing towards the cluster. Such tendency becomes more important when related to the study of the Virgo morphology by West & Blakeslee (2000), who determined a triaxial structure. The distribution of Virgo dwarf elliptical galaxies seems to elongate to the northern part (Binggeli et al., 1987; Binggeli, 1999) as does the distribution of hot X-ray -emitting intracluster gas (Schindler, Binggeli & Boheringer, 1999). In fact, the line connecting M87 and M84 appears as a main axis of the cluster. Considering only the BCG of Virgo, these authors traced a highly elongated, three-dimensional structure that forms part of, and is oriented parallel to, a larger filament passing through the heart of Virgo. On the theoretical side, numerical simulations were performed by Hatton & Ninin (2001), who searched for mutual alignment of angular momentum vectors of dark halos, and alignment with structures, as a function of the distance with negative results. But in this case, the lack of correlation seems to be related to the small number of halos used. On the contrary, they detected a strong signal for the alignment of haloes with the surrounding halo distribution on scales $1 - 30h^{-1}$ Mpc. Bond (1987b) proposed that these tails are an intrinsic property of the primordial density peaks.

Some enlightening numerical simulations have given important information on the kind of cosmological scenarios in which these anisotropies are produced. For example, West, Dekel & Oemler (1989), carried out several simulations to reproduce the anisotropy and to test the hypothesis that it can rule out some models of structure formation. Their study included: a) a pancake scenario with spectral index $n = 0$, b) an hybrid scenario, and c) three hierarchical scenarios with power spectrum indices $n = 0, -1, -2$ respectively. In the pancake scenario, it was found a correlation between the orientation of clusters with respect to the galaxy distribution on scales up to at least $10h^{-1}$ Mpc. Somewhat weaker evidence of alignments was found for the rest of the models, with the $n = 0$ hierarchical model showing the weaker effect. This result could reflect the fact that the small-scale fluctuations collapse much before the cluster and supercluster formation without any connection. On the contrary, the positive results of an anisotropic distribution of galaxies, for the cases $n = -1, -2$, suggest that there is an important contribution of the large-scale density field to the density peaks, which should make that collapse of density peaks at different scales occurs almost simultaneously. In such a case, any anisotropy in the shape of the peaks and their possible connections will be increased.

Let us now address some words to the case of gravitational lensing. This effect is a powerful method to trace directly the distribution of mass in the Universe, as the gravitational potential produced by luminous and dark matter induces coherent distortions on the observed shapes of background galaxies, which, most of the time, are far away from each other in redshift space. Therefore, a logical hypothesis to adopt, in order to model the lensing effect, is a distribution of galaxies uncorrelated in orientations and ellipticities. However, the correlations in ellipticities and orientations induced by lensing requires an intrinsic correlation, typically $\sim 10^{-3} - 10^{-4}$ indicating that intrinsic alignments cannot be easily neglected (*e.g.*, Heavens et al., 2000; Heymans & Heavens 2002; Pen et al. 2000; Croft & Metzler 2000; Catelan et al. 2001; Crittenden et al. 2001, 2002). Some efforts for example have been done by Crittenden et al., 2001 and Mackey et al., (2002), to separate the intrinsic correlation from a weak lensing signal by measuring E-mode and B-mode correlations (*e.g.*, Miralles et al., 2002). However, the main drawback of this method is the need of an accurate correlation between these two modes (Schneider, van Waerbeke, & Mellier, 2002), not currently available.

Large numerical simulations released by the Virgo Consortium, were used by Heavens et al., (2000) and Croft & Metzler (2000) to measure the ellipticity correlation function c_{ij}

$$c_{ij} = < \epsilon_i(x)\epsilon_j(x+r) >,$$

for dark matter halos in weak lensing experiments. The main assumption done is that ellipticity correlation of galaxies shall have the ellipticity correlation of halos. Their analysis suggests that intrinsic correlations of galaxy ellipticities can really dominate over the lensing signal in shallow surveys, whereas this could be a non-negligible signal, as high as 20 percent, in deep surveys. The main worry with these simulations is that correlations vary according with the resolution, and with the slightly different criteria for the halo selection. Nevertheless, after correcting these problems, Jing (2002) probed that intrinsic ellipticity correlations converge reasonable well in halos of big $L = 300h^{-1}$ Mpc and small $L = 100h^{-1}$ Mpc boxes, meaning that it can contribute to contaminate not only to shallow surveys but also to deep surveys.

Working out an analytical expression for the ellipticity (or orientation) correlation is not trivial, whereas numerically the problem is resolution and therefore it would be hard to get an important signal-to-noise. But, some analytical work has been done by Catelan & Porciani (2001) and Catelan et al.,(2001). Given the Lagrangian position $\mathbf{q}$ of a density fluctuation, δ, one can calculate the gravitational potential via the Poisson equation, thus it is easy to get the deformation tensor produced by the surrounding distribution of matter as

$$D_{\alpha\beta} \equiv \partial_\alpha \partial_\beta \phi,$$

where the tidal tensor is in fact defined in terms of the gravitational potential ϕ as

$$F_{\alpha\beta} = \left(\partial_\alpha \partial_\beta - \frac{1}{3}\delta_{\alpha\beta}\nabla^2 \right) \phi.$$

According to these authors, once the tidal-tidal correlation is calculated, a term appears that makes evident an anisotropy in this correlation, *i.e.*, with respect the separation, orthogonal tidal correlations show the tendency to be stronger than the longitudinal tidal correlations.

In spite of these conclusions, these results are not an irrefutable probe of their suitability, as secondary infalls and strong and frequent mergers could change the results.

5. The Mechanisms Affecting Orientations

5.1. Primordial Origin for the Preferred Orientations

This section is devoted to the present arguments that are in favor of a primordial origin for the alignment effects.

At the early stages of the structure formation epoch, density perturbations gravitationally interacted and were subject to the influence of tidal fields, which are generated by the anisotropic distribution of neighboring perturbations. The subsequent response and evolution of density peaks to the tidal field mainly depend on:

1. the shape of the perturbation;
2. the strength of the external tidal field;
3. the relative orientation between the shape and the field.

Concerning shapes, the triaxial morphology of peaks in Gaussian random fields has been fully confirmed (Peacock & Heavens, 1985; Bardeen et al., 1986; González, 1995; 1995a). The tidal field, then, gets coupled with the quadrupole moment of the perturbation, exerting on it a torque and imprinting it with angular momentum. The torque per unit mass at a distance R from the perturbing body decreases as R^{-3}, but increases with the mass. If the power spectrum of perturbations is flat on large scales (Vogeley et al., 1992; Peacock & West, 1992; Einasto et al., 1993; Bernardeau et al., 1997; Falthebacher et al., 2002), then the mass increases as R^3, and therefore a divergence in the total torque could in principle appear. This suggests that the local properties of density fluctuations can be affected by the large-scale density field (Barnes 1992; Pen et al. 2000; Catelan et al., 2001).

The orientation of incipient perturbations is also of great importance because it is closely related to the amount of angular momentum to be acquired. Density perturbations that are born with their major axes aligned with those of the tidal tensor experience a null or weak torque, and therefore would acquire a low amount of angular momentum. At least there is a secondary infall of material. It is possible, in principle, that some of these perturbations preserve their initial orientations even after the non-linear evolution of the density field. Whether (or not) the initial number of these peaks is statistically significant is a subject that has not yet been studied in detail. There is increasing evidence, both observational and theoretical, which suggests that this number could be important in the past and is preserved today.

A primordial origin for alignment effects has been suggested by several authors *e.g.* Barnes & Efstathiou (1987); Lambas, Groth, & Peebles (1988b); Bond (1986, 1987a,b,c); West (1994); González (1995, 1995a). Bond (1987) and Bond et al., (1996) for example, hit on the idea that Gaussian 'cross-talk' can produce quite large anisotropies in the field around cluster-scale density peaks; *i.e.*, those density waves that constructively superpose to build the peak, also preferentially constructively interfere along the long axis compared with the other two directions. Thus, this coherent wave bunching would tend to shape the perturbations largest along the long axis and smallest along the short axes so that they point toward each other. Pauls & Mellot (1995) have shown how, even in simulations with abundance of small-scale fluctuations a pancake-like collapse associated with long-wavelength modes still determine the final positions and orientations of objects. If this is right, the orientation of a central, smaller-scale peak would be correlated with the surrounding cluster-scale orientation, especially if the two scales are not exaggeratedly different. As a natural consequence of these anisotropies, cluster-scale perturbations would tend to align with their surroundings. Moreover, in the N-body simulations carried out by Barnes & Efstathiou (1987), it was found that the alignment of the major axes of the formed objects is more prominent than is the alignment of angular momentum vectors of nearby objects. As they pointed out "*[...] The most striking effect seen in these tests is the tendency for nearby objects to point at each other. It seems likely that objects are born with these orientations; if they had been formed with random major axes and sheared into line by tidal torques, we would expect a strong spin-vector versus separation-vector effect, which we do not find*". If tidal torques between neighboring structures have not been a determining factor in the generation of alignments, then how do these coherent patterns arise?

Yet the possibility that the wide gamut of alignment effects on different scales might have a common, primordial origin has not been studied in detail to date. This is the subject we will address next. For this goal, in the rest of this section, I present a short historical review of the subject. In Section 2, we briefly discuss the evidence for small-scale alignment effects and the possible physical mechanisms involved. Observational and theoretical evidence for several large-scale alignment effects is presented in Section 3. In Section 4, we describe the theoretical evidence in favor of a primordial origin for a number of alignment effects. Finally, in Section 5 we present our conclusions.

5.2. Alignments in HDM and CDM Models

Observations of alignment effects date back to a time long before the fundamentals of the cosmological scenarios, presently in vogue, were completely set down (Abbe 1875; Knox 1908; Meyermann 1923; Gregory 1924; Brown 1939, 1964, 1968; Wyatt & Brown 1955). These authors analyzed the spin vector of galaxies in the Local Supercluster (LSC)(*although they were not aware of its existence!*), finding that the spin vectors tend to cluster around a particular plane which lies close to the plane of the Milky Way and is almost perpendicular to the plane of the LSC. However, the first physical attempt to explain this feature of galaxies was elaborated by Reinhardt (1971) in terms of a sort of coupling of the angular momentum vector of galaxies to a large-scale primordial magnetic field. Consequently, Reinhardt recognized the need for a physical mechanism, such as the magnetic field, homogeneous on large-scales but inhomogeneous on scales of galaxies and clusters.

The suspected network of interconnected filamentary structures was soon detected by two opposite theories of structure formation: the Hot Dark Matter (HDM) and the Cold (CDM) Dark Matter model. Under the framework of the HDM model [*e.g.* Doroshkevich, (1970; 1973a,b); Doroshkevich, & Shandarin (1978); Doroshkevich, Shandarin, & Saar (1978)] filaments and alignments arise together as a natural consequence of the fact that the spectrum of density fluctuations possesses a coherent length associated with the scale of neutrino free-streaming, which damps out small-scale perturbations below a critical wavelength. Clusters and superclusters are the first structures to form, evolving according with the Zel'dovich approximation (1970); later on, galaxies form by fragmentation of the collapsing parent pancakes. If the fragmentation process is delayed until the late stages of the collapse, then it is expected that the orientation of the major axes of the fragments are parallel to the pancake plane. For a collapse highly anisotropic, occurring in two directions, the orientation of the fragments remember the direction of the filament.

Once a pancake enters the nonlinear stage, gas collisions induce the formation and propagation of shock waves through the pancake, which is thus compressed and heated. During the linear phase of growth, perturbations have zero vorticity, but now shock waves increases the gas entropy and violates vorticity conservation. The velocity component, normal to the shock wave vanishes while the tangential component remains unchanged, thus the vorticity is generated parallel to the pancake (Dekel, 1985). Subsequently the pancake undergoes radiative cooling and fragments, so that clusters and galaxies form. The final results of this process are (Doroshkevich, 1970):

1. The angular momentum of spiral galaxies must be tangent to the plane of the pancake. Doroshkevich (1970; 1973a,b) and Binney (1974) argue that before the dissipation occurs, the regions which eventually collapse to a galaxy could have a small amount of angular momentum. If the main axes of these regions are parallel to the main axis of the pancake, the mean angular momentum vanishes. However, its dispersion is zero only along the axes of the protogalactic region. Conversely, if the axes do not coincide the mean angular momentum is non-zero.

2. The vorticity increases with the distance from the center of the pancake: hence elliptical galaxies – which have low rotation velocity – must have been formed in the central regions of aspherical clusters and superclusters, where the vorticity is small.

3. The vorticity diminishes with the pancake mass: hence a majority of elliptical galaxies must exist in larger clusters. The elliptical galaxies, whose shapes are determined by the initial anisotropies in the velocity dispersion rather than by rotation, are expected to be aligned with their parent pancakes. Every element of fluid in the pancake is deformed due to the growth of the perturbation itself and by the expansion of the universe.

Doroshkevich (1970) has shown that the principal axes of the tensor associated with these deformations, are oriented in similar directions at every point of the pancake. These directions define the pancake plane in which the alignment of elliptical galaxies occurs. The numerical simulations of Elmegreen & Elmegreen (1978) and Aarseth & Binney (1978) – which showed that the fragments of thin pancakes end up aligned along the pancake plane – support this idea. The shape of the pancake and the number of fragments produced is important to retain or erase away any signal of the coherent pattern of alignments. This is because, as soon as these fragments develop in clusters and galaxies, they start to experience tidal interactions and mergers which could modify their initial orientation. The HDM scenario, however, has been ruled out as it presents serious drawbacks (e.g., Frenk 1990): late galaxy formation, excessive clustering of matter, a transient character of filaments, etc.).

Initially, hierarchical scenarios were thought to be unable to account for alignments primarily because tidal torques are short-range interactions decaying with distance at r^{-3}.

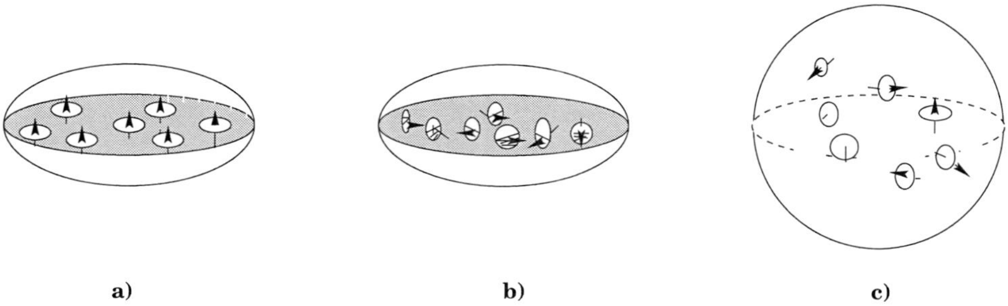

Figure 3. The orientation of galaxy spin as predicted in three models. *a)* Primordial vorticity scheme, *b)* pancake shock scheme, and *c)* tidal torque scheme. From Iye & Sugai (1991).

Therefore, they could not account for cluster-cluster alignments observed on scales up to $30 - 60h^{-1}$ Mpc (Binggeli, 1982; Rhee & Katgert, 1987; West, 1989; Lambas et al., 1990; Plionis, 1989, 1994; Plionis et al., 1992; West & Schombert, 1993). Second, if tidal interactions were responsible for the acquisition of angular momentum and alignments, a correlation would be expected between the orientation of the spin-vectors and the distance to neighboring structures. Such a kind of correlations have not been found observationally nor even in N-body simulations (Barnes & Efstathiou, 1987). In fact, positive alignments could be diluted if a large sample of galaxies in cluster, or larger structures, is considered (Kitzbichler & Saurer, 2003, Sugai & Iye, 1995). At the light of the weak lensing detection of intrinsic galaxy alignments, large-scale tidal interactions are now revisited as the favorite candidate to produce alignments (Catelan & Porciani, 2001; Lee & Pen, 2000a,b; Catelan & Theuns, 1996; van de Weygaert, 1997; Salvador-Solé & Solanes, 1993; Binney & Silk, 1979). Nevertheless, Coutts (1996) used circular statistics to quantify the effect of random tidal torques on galaxies, and determined that an initial alignment is not greatly dispersed by torques and should remain detectable for small samples. In fact, in order to settle non-random orientations on a basis of physical reality, some work has done to rule out projection effects by MacGillivray et al. (1980) performing Monte Carlo simulations of a number of galaxy distributions.

N-body simulations have already probed the ability of the CDM model to generate both filamentary structures (e.g., Efstathiou et al., 1985; Efstathiou, 1990; Frenk et al., 1988) and cluster-cluster alignments extending to $\sim 30h^{-1}$ (West, Villumsen, & Dekel, 1991; Kofman et al., 1992; Bond, Kofman, & Pogosyan, 1996, Onoura & Thomas, 2000). It seems that the major requirement to generate filaments and alignments is a sufficiently flat power spectrum at large-scales.

One main assumption in the structure formation process is that galaxies and clusters of galaxies might have formed at the site of high density peaks in the primordial density field. Following this idea, Kaiser (1998) proposed that rich Abell clusters, which are the most massive systems collapsing today, could be formed from rare $\geq 3\sigma$ fluctuations. This can account for the stronger clustering of Abell clusters compared to that of galaxies, since rare high-σ peaks have a statistically enhanced clustering tendency relative to the underlying mass distribution in Gaussian random fields.

Several authors (e.g., Blumenthal et al., 1984; Dekel, West, & Aarseth, 1984; Silk &

Evrard, 1988; Evrard, 1989; Evrard, Silk, & Szalay, 1990) have also proposed that galaxy type might be determined by the peak height, with early-type galaxies forming from $\geq 3\sigma$ peaks, whereas late-type galaxies might be formed from $1-2\sigma$ peaks. Thus, if normal bright elliptical galaxies were formed from rare peaks, then the most massive galaxies (BCG's), generally located at the centers of rich clusters, must have formed in these special sites under a special process as proposed by van den Bergh (1983), Sandage (1976), and Tremaine & Richstone (1977). One of these special processes may be the anisotropic merger mechanism explored by West (1994), who was motivated by the systematic higher galaxy counts out to distances $\simeq 15h^{-1}$ Mpc from the centers of Abell clusters in a direction defined by the major axis of BCG.

5.3. Cluster-Cluster Alignment

In the numerical simulations carried out by West, Villumsen & Dekel (1991) and West, Dekel & Oemler (1989), they found that cluster alignments occur only when the initial fluctuation spectrum possesses a large coherence length somehow, as that expected in the HDM model. On the contrary, hierarchical clustering models are unable to reproduce the observed alignments. In these models, however, tidal effects could influence the shapes of protoclusters and induce alignments. Yet, when Binney & Silk (1979) discovered that tidal effects between protoclusters can induce flattening and prolate shapes for clusters, Barnes and Efstathiou (1987) using numerical simulations found that tidal interactions cannot produce alignment between neighboring protoclusters, and with smaller probability, alignment on scales up to $\sim 50 - 60h^{-1}$ Mpc (West, 1989). On the contrary, by using analytical methods, Salvador-Solé & Solanes (1993) found that alignment of clusters can be produced by tidal fields. Contrary to these conflicting results and to the conclusions by Dekel et al., (1984); West et al., (1989), and Bond (1987a,b,c), was the first to show that random Gaussian density fields are able to produce alignment of clusters up to $\sim 20 - 30h^{-1}$ Mpc if they form at the highest peaks of the field.

Aiming to elucidate the origin of the alignment of clusters of galaxies and whether they could be used as a constraint on the cosmological scenarios of structure formation, West, Villumsen & Dekel (1991) performed an analysis of the alignment of clusters by using N-body simulations. Their study covered a wide range of models: featureless power law spectra, the CDM, HDM, and Mixed DM scenarios, and the explosion scenario as well as different values of the density parameter Ω. They got two important conclusions: first, the existence of alignment for clusters of galaxies is a natural outcome in several cosmological models, and therefore that those effects cannot be used to *exactly* constrain cosmological models. Second, a) a pancake scenario with $n = 0$ on large scales and no fluctuations below some critical coherence wavelength; b) a CDM model with $n = -2$ more appropriate on scale of galaxies; and, c) a hybrid scenario with $n = 0$, can *produce* significant alignment of galaxy clusters. All these models require a universe with $\Omega = 1$. A trivial but important aspect is that the alignments of clusters were detected at the end of a Hubble time, which was fixed in the simulations by Dekel & West when the slope of the 2-point correlation function matched the current slope.

Nevertheless, several cosmological N-body simulations (Efstathiou, Davis, Frenk & White, 1985, 1986a-d, 1987a,b; Melott & Shandarin, 1989; West, Villumsen, & Dekel,

1991; Kofman et al., 1992) have shown that the development of filamentary structures is a generic feature of many cosmogonic scenarios possessing a sufficiently flat spectrum of density perturbations. West (1994) has pointed out that if the initial power spectrum has the structureless scale-free form $P(k) \propto k^n$, then the criterion for the development of filamentary and sheet-like structures is approximately $n \leq -1$, with $n \approx -1$ marking the critical transition between cosmogonies driven primarily by the formation of a hierarchy of roughly spherical clumps and those characterized by the emergence of large-scale anisotropic features. Flatter power spectra lead to a more pronounced filamentary appearance. Furthermore, West (1989) and Plionis et al., (1992) in their study of clustering of galaxy clusters found evidence of correlation between the shape of superclusters and the cluster-cluster alignment effect, with the alignment being more significant in superclusters which are nearly prolate.

Once one compares the type of models which develop filamentary structures with the models determined by West, Villumsen, & Dekel, (1991) presenting alignment effects, it is possible to set down a working hypothesis which has been explored already for some time (González, 1995; Bond, Kofman & Pogosyan, 1996): these two aspects are intrinsically correlated and produced in models characterized by flat spectra.

Whereas several N-body simulations have probed the ability of a number of cosmological scenarios to reproduce cluster-cluster alignments and alignment of cD galaxies with the major axes of their host cluster at the end of a Hubble time, very few have been carried out to analyze their origin or possible evolution. In what follows, we will review some important evidence addressed to propose the idea that these alignments do exist in the initial conditions. In particular, we will review the hypothesis that the small-scale density peaks are oriented in the same direction as that imposed by the larger scale density peaks as a result of the dominant feature of the long density waves of the field. The central idea is that the constraint imposed by requiring a cluster to be oriented along some direction, specified by the major axis, can force the density waves – which had to constructively interfere at the peak – to also preferentially constructively interfere along the long axis compared with other directions.

6. Evolution of alignments

6.1. Dilution of Alignment Signals by Successive Close Encounters

As we have alluded to, anisotropies in the orientation of galaxies are closely related to the intrinsic aspherical shape of their parent structures. However, in order to explore its origin as the result of an evolutionary process in clusters of galaxies, it is essential to analyze the mechanisms that could establish or destroy such a coherent pattern. This section is devoted to the analysis of tidal interactions with the host cluster potential and dynamical friction as one of the mechanisms.

The first idea proposed to explain such alignment within the context of hierarchical models, is that close neighbor galaxies can interact by the means of tidal forces. Such galaxies are expected to line up with their planes parallel but with the spin vector antiparallel. This mechanism would satisfy the alignments observed for galaxies of type Sab-Sc in a flattened system; however it is unlikely to be able to account for the preference observed for winding directions since equal numbers of "**s**" and "**z**" (opposed to reversed **s**) types are to be expected.

Wesson (1982, 1984, 1985) has studied the influence of a cluster tidal torque acting in spiral galaxies and concludes that, because of the angular momentum, their orientations are stable over times greater than a Hubble time. Also for spherical galaxies with little or null angular momentum, such as an elliptical, their orientations are unstable on a time scale comparable with a Hubble time. Hence, such a mechanism cannot produce alignments in rich clusters. Instead of a tidal mechanism, MacGillivray & Dood (1985) proposed that the agency responsible for the formation of the LSC acted only in a weak manner to produce the alignment effects observed. They argued that in the contracting, low-density proto-LSC, the first galaxies to form were the progenitors of the current Sab-Sc spirals, and that the primordial effects are more strongly reflected in the properties of this galaxies. Subsequently, as the LSC continued its collapse, other mechanisms – *e.g.*, tidal or gravitational interactions – became more important, gradually erasing any memory of the primordial orientations.

A suggestive point can be inferred when we compare the results of some of these analyses. A characteristic which is common to all the clusters of the sample analyzed by Adams et al., (1980) is that – with the exception of A2197 – they are of the richness class 0. This means that on average the density of galaxies in the clusters is low, and therefore interactions with neighboring galaxies are less frequent. If this is the case, there may not have been sufficient encounters to modify the initial orientation distribution of galaxies (*e.g.*, Coutts, 1996). Consequently, the detected parallel alignments would point to an intrinsic origin of galaxy alignments related to the aspherical shape of their parent cluster. The clusters used by Thompson, on the other hand, include Coma, Hercules, and Virgo which have richness ≥ 1. From these, Coma was the only cluster that exhibited radial alignments.

Some insight on the answer to this question can be gained by estimating the average number, N_e, of strong encounters that a galaxy experiences in a Hubble time, in clusters of different galaxy number density, n. This is given (Binney & Tremaine 1987) by

$$N_e = \frac{3\sqrt{2}}{4} vb^2nt, \qquad (2)$$

where v is the galaxy velocity, $b = 2GM/v^2$ is the impact parameter deduced from the virial theorem and M is the mass of the galaxy. We can obtain an upper limit to N_e by considering typical values for the Coma cluster. If we take $v = 800$ km s^{-1}, $b \approx 10$ kpc, a total number of galaxies $n_g \approx 5 \times 10^3$ in a region with a mean radius of $R \approx 2.5$ Mpc, then the number density of galaxies $n = n_g/(4\pi R^3/3) \approx 76$ Mpc^{-3} gives us $N_e \approx 0.061$. Thus, about 6.1% of the galaxies in Coma could in principle have experienced interpenetrating collisions (Thompson, 1976). In denser clusters, or equivalently in the core of the clusters, the number density of galaxies is higher and therefore a larger number of collisions should occur. Thompson & Gregory (1978) suggested that the radial alignment trend observed in Coma might be produced by distortions in the shape of galaxies traveling on radial orbits through the core of the cluster. Because of a high frequency collisions, galaxies might become extended in their direction of motion, *i.e.*, the radial direction, making them appear to point to the cluster center. This possibility has been studied by Ciotti & Dutta (1994) who considered the isophotal distortions of elliptical galaxies moving in either radial or circular orbits. In this picture, the tidal force produced by the cluster center is more important than the tidal field produced by neighboring individual galaxies.

Galaxies within non-spherical clusters of richness 0 experience a low number of encounters. In this case, the number density of galaxies is $n \approx 20$ Mpc^{-3}, which gives an average number of collisions per galaxy $N_e = 0.035$ – *i.e.* about $3 - 4\%$ of the galaxies could have strongly interacted. Therefore, encounters of galaxies would be unable to destroy any initial alignment effect. It is noteworthy then, that in three clusters of richness 0 of the eight clusters analyzed by Adams et al., (1980) a significant parallel alignment is found.

We can go to even a lower galaxy number for densities, when at the level of groups. In this case, the cross-section of the galaxies is larger than in high density environments due to the low velocity of galaxies and therefore the collisions are more frequent. By taking a typical velocity $v = 200$ km/s, $b \approx 100$ kpc, $n_g \approx 10^2$ and mean radius of 0.7 Mpc the number of collisions is $N_e \approx 0.29$. Thus a large fraction, 30%, of the galaxies could have collided, so that the information on the initial conditions prevalent at the epoch of their formation would have been erased. Alignments should not be present in groups. Nevertheless, dominant galaxies in poor clusters seems to show the tendency to be aligned with the group mayor axis (Fuller, West, & Bridges, 1999). The source of these contradictory results seems to be the isotropy in the interactions. Anisotropic mergers of material forming clusters could produce coherent patterns.

There are some other observational studies which suggest the importance of tidal torques in the generation of alignments. One of these comes from the study of the angular momentum vectors of adjacent galaxies. A first study by Sharp et al., (1979) showed an isotropic distribution of the position angles between the major axes of spiral galaxies in pairs. However, Helou (1984), reported the presence of strong correlations of anti-parallel alignment of the angular momentum vectors in a sample of 31 binary spiral galaxies (see also Flin, 1993). In the same context, Oosterloo (1988) divided this sample into 20 isolated pairs and 11 pairs in groups finding evidences that the anti-parallel alignment in the distribution of the angular momentum vectors is stronger for the pairs in groups. Other analysis on the projected direction of the rotation axes of interacting galaxies showed that the pairs tend to be orthogonal to each other (Sofue, 1992), a result confirmed by Muriel & Lambas (1992), who also found a preferred orientation of the major axis of spiral galaxies in the direction toward their nearest neighbor on scales of $3h^{-1}$ Mpc. Elliptical galaxies were found to be preferentially oriented not only with their nearest neighbor but also with all neighbors on scales of $3h^{-1}$ Mpc.

A problem which is closely related to that of the orientation of galaxies is the acquisition of angular momentum by density perturbations. The most popular idea on its origin is that angular momentum is acquired through tidal torques between neighboring perturbations. Much work has been done to quantify the angular momentum gained by the perturbations (e.g., Peebles 1969; Doroshkevich 1970; Efstathiou & Jones 1979; White 1984; Barnes & Efstathiou 1987; Heavens & Peacock 1988; Caimmi 1989; Quinn & Binney 1992; Voglis & Hiotelis, 1989; Ueda, Itoh, & Suto, 1993, and Ueda et al., 1994). For the purposes of this paper, there are some relevant results in the numerical simulations by Barnes & Efstathiou (1987). In order to quantify the acquired angular momentum of perturbations they considered three sets of models: a white-noise ($n = 0$) model, and two with more power ($n \leq -1$) on large scales. The acquired angular momentum, however, was independent of the spectral index of the density field spectrum. They confirmed that the angular momentum and the major axes of the formed objects tend to be perpendicular and also that nearby objects tend to point at each other both in CDM and white-noise models. Furthermore, in all of the models a coherent field was found, in which the major axes of groups of particles are contained. This result becomes more surprising because of the fact that there was no evidence of a correlation of angular momentum vectors between neighboring objects of the kind reported by Helou (1984), Oosterlo (1988), Muriel & Lambas (1992), and Sofue (1992). If the perturbation population possesses an isotropic distribution of orientations, and later becomes aligned by the action of tidal forces, then a strong spin-vector versus separation effect should be detected; because of angular momentum conservation, isolated pairs of objects can exhibit anti-parallel spin-vectors. Given that this effect was not seen, small-scale tidal forces should not produce important orientation effects. The alternative is, as prompted by Barnes and Efstathiou, that those objects are born with these orientations.

6.2. Formation of cD Galaxies

These kinds of coherence (see Figure 3), has not only important repercussions for our theories of cD formation, but also on the coherence of filamentary structures to scales probably larger than $40h^{-1}$ Mpc (see the next discussion).

Till very recently, the most popular theory regarding cD galaxy formation was that they formed via special dynamical processes either during the collapse of the cluster (Merritt, 1983, 1984, 1985) or after it had virialized (Sandage 1976; Ostriker & Hausman 1977). Any normal galaxy that spent a significant fraction of its dynamical time away from the cluster center may become tidally stripped by the cluster tidal field. So, in the post-virialization model, cDs started out as "normal" elliptical galaxies located at the dynamical centers of rich clusters, and grew by the accretion of material stripped from other galaxies. Also, the dynamical friction between galaxies and the smooth intracluster medium dragged the larger galaxies to the center of the cluster where they merged with the primeval cD. In this theory it is considered feasible that massive galaxies exist in the innermost region of clusters. Sooner or later the alignment of the cD major axis and the cluster would arise because the accreted halo material reflects the characteristic shape and orientation of the cluster potential (*e.g.*, Rhee & Roos, 1990; Garijo, Athanassoula, & García-Gómez, 1997).

Three main predictions result from this theory: the cD galaxy must become oriented with the cluster potential; it must not only be located in the center of the cluster, but also it must be at rest with respect to the cluster. Malamuth, Smith, & Kriss (1989) examined the X-ray emission of clusters, and found that while the peak of this emission coincides with the cD galaxy, the outer isophotes are not centered on the cD. In fact, some clusters show evidence for subsystems of galaxies which have not reached virial equilibrium with the rest of the galaxies (Fitchett 1988). Lauer (1988, 1990) provided some evidence for multiple nuclei of BCG, which seems to indicate that cDs may be evolving. If these results receive further confirmation, one could talk about the orientation of the inner region of a cD galaxy with the parent cluster, but not about the orientation of the whole galaxy. Wesson (1982, 1984, 1985) and González (1995a) by using an over-simplified quadrupole model of a galaxy – which in the presence of the cluster tidal field and dynamical friction, behaves as a damped harmonic oscillator – estimated the time necessary to become aligned with the cluster major axis, the result being of the order of a Hubble time. The merging process is not instantaneous, *i.e.*, the time necessary to form the cD galaxy and then to align it with the cluster would imply that those galaxies are in the process of formation and therefore be aligned. cD galaxies should, therefore, be observed with random orientations. The evidence, however, seems to point in the direction that they are evolving, not forming.

Because of the isotropy of the merging process, cD galaxies are expected to form with low peculiar velocities. This prediction is in contradiction with the observational measures of the peculiar velocity by Sharpley, Ellis, & Gray (1988), Hill et al., (1988), and Malamuth (1992) of $\delta v \sim 400 \text{kms}^{-1}$ with respect to the cluster mean. Hill et al., (1988) for example, tested the hypothesis that cDs are formed in the virialized cluster and estimate their peculiar velocity either from two-body interactions or as the residual momentum from the galaxies that merge. They find that in order to reproduce their high peculiar velocity, cDs must be no larger than 4-5 times as massive as an average cluster galaxy, or have had fewer than three mergers. In the study by Lauer (1990) of the surface brightness of multiple nuclei BCGs, he finds that only about 25% of the multiple nuclei systems are mergers in progress. He assumed that a secondary nucleus remains for about one orbital period, $\sim 3 \times 10^8$ yrs, and that the merger rate over the age of the cluster is the same as the present merger rate. He finds that on average a BCG will grow by approximately $4L^*$ or about one-third of a typical cD luminosity. Thus, his main conclusion is that mergers in the present clusters are an important ingredient in the evolution of cD galaxies, but are not sufficient to form them. The difficulty of forming giant $\sim 10^{12} M_{\odot}$, galaxies in a Hubble time, by adopting the present day merging and accretion rates, has also been confirmed by Merritt (1985) and Tremaine (1990).

There are, some other drawbacks of the isotropic merger model of cD formation. BCG luminosities have a small dispersion, which it would be difficult to explain since they are formed in clusters where a number of different environments prevail – at least if they formed by cannibalism and accretion in a sophisticated self-regulating process. Is is also difficult to explain why the innermost $\sim 2h^{-1}$ kpc regions of BCGs should be aligned with their surroundings: if they started out as normal elliptical galaxies embedded in an accreted halo, then like other normal elliptical galaxies their cores should be randomly oriented.

Another aspect closely related with the tendency of BCGs to be distributed in a linear pattern along the cluster mayor axis, is the correlation with their surrounding mass distribution. A number of studies (e.g., Binggeli, 1982; Rhee & Katgert, 1987; West, 1989, 1994; Plionis et al., 1992; Rhee, van Haarlem, & Katgert, 1992; West & Schombert, 1993) have shown that the major axis of Abell clusters exhibits a prominent tendency to point towards neighboring clusters over distances of up to several tens of Mpc. These results become even more suggestive of a large-scale coherence when one considers the high anisotropy in galaxy counts along the major axis of clusters out to separations of $\sim 15h^{-1}$ Mpc from the centers of Abell clusters (Argyres et al., 1986; Lambas, Groth, & Peebles, 1988a,b). Figure 4 shows evidence of the connection between the structure of BCGs and the surrounding galaxy distribution. The figure compares the projected major axis orientations of the innermost regions of a sample of 147 BCGs with the distribution of all neighboring Abell clusters within a distance of $\sim 10h^{-1}$ Mpc. The angle θ is the difference between the PA of the BCG major axis and the PA between the galaxy and a neighboring cluster. The θ frequency distribution shows that the orientations of the innermost BCG regions are correlated

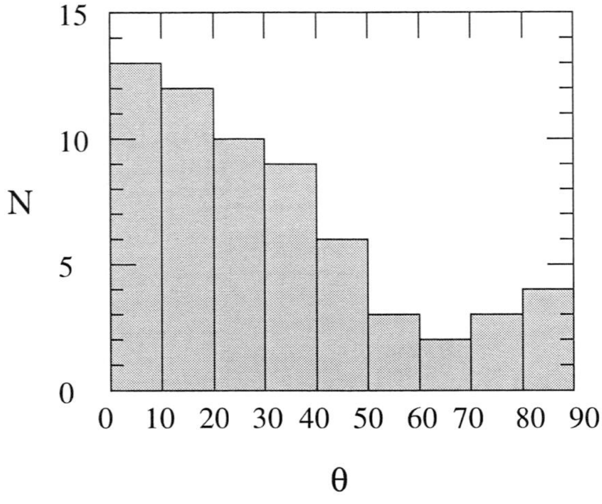

Figure 4. Alignment of the innermost regions of a sample of 147 BCGs with the large-scale environs. θ is the angle between the projected BCG major axis at a radius of $\sim 2h^{-1}$ kpc and the great circle between the BCG and another neighboring Abell cluster within $10h^{-1}$ Mpc. The significance of the alignment effect exceeds 99.9% .

with their surroundings.

X-ray and radio-based surveys have also detected a coherence in the BCG-Cluster orientations. Allen et al., (1995), for example, combining the data of the PSPC (Position Sensitive Proportional Counter) of ROSAT with optical CCD images and spectra of five of the most luminous, intermediate redshift clusters, detected a positive relationship only for A1068, A1361, and A1664. However, the PSPC X-ray data indicated that for these cases, BCG lie at the centers of massive cooling flows (200-400 $M_{\odot}$/yr). For those clusters with non-important cooling flows, the coherence seems to be negligible. The advantage of using X-ray studies over optical ones is the determination of the collision time-scale of the intra-cluster gas, which is shorter than its dynamical and cooling time-scales. Thus, even when the galaxy distribution depends on subcluster merging, occurring after the initial cluster collapse, the X-ray intracluster medium provides a tracer of the evolving cluster potential. A special mechanism must, then, be invoked to transfer the morphology information of the cluster at a scale of Mpc, to the BCG in the innermost few Kpc. An insight into this suggestion is the detection of wide-angle tails of radio sources "blowing" towards the main axes of superclusters Novikov et al., (1999).

West (1994) made a "novel" suggestion taken from van den Bergh (1983) to propose a substantially new theory for the formation of cD galaxies which accounts for some of the drawbacks of the standard theory.

6.3. The Effect of Anisotropic Mergers

Van den Bergh (1983) suggested that cD galaxies were special at the early stages in the evolution of their parent clusters, their sites being where clusters formed.

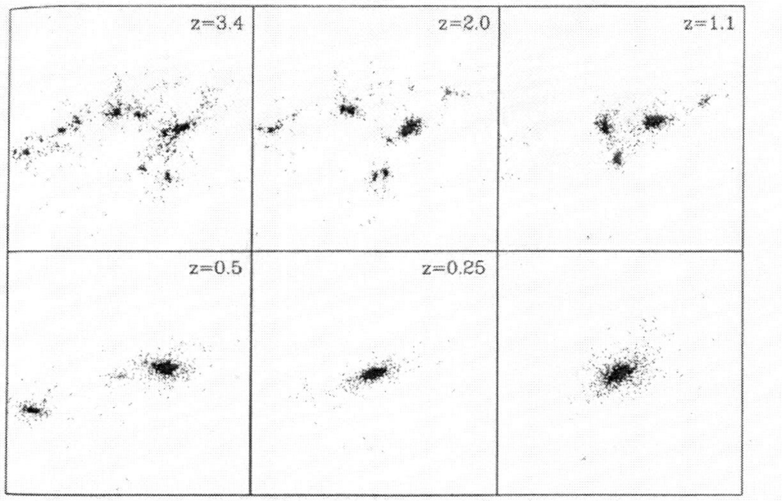

Figure 5. N-body simulation showing the formation, via anisotropic mergers, of a massive cD galaxy in the center of a rich cluster. Note the orientation of the cD galaxy at low and high redshifts. From Carlberg (1993).

Following this idea, West (1994) proposed that cD galaxies were formed via anisotropic mergers of substructures moving toward peaks – the special sites proposed by van den Bergh (1983) (see also Sandage, 1976; Tremaine & Richstone, 1977). The fundamental ingredient in this theory is that the merging objects move along the main axes of the peaks which are oriented parallel to large-scale anisotropies in the primordial density field. At the center of these rich protoclusters, the deformation tensor of the local density field is greatest, and therefore the material flowing along developing filamentary structures will accumulate in the potential wells of peaks. Thus, the peaks are the convergence sites of the infall material, gas and large substructures which follow the direction and orientation (Godlowski & Baier, 1997; Godlowski, Baier, & MacGillivray, 1998) of the large-scale filament.

This theory is supported by the results of the one million particle simulations by Carlberg (1993) of a rich Coma-like cluster in CDM models, where it was observed that the formation of the cD galaxy proceeds through the mergers of a number of objects, rather than from the collapse of a single density perturbation (Figure 5). Additionally, mergers of these objects occurred in a direction dictated by a developing large-scale filamentary structure. Similar effects have also been found by Mellot & Shandarin (1989), van Haarlem & van de Weygaert (1993) and Dubinski (1992, 1998), who noted an anisotropic flow of material falling towards big clumps of matter distributed along the filamentary structures. As a result of this formation mechanism, the cD galaxy should retain some memory of the anisotropic process from which was created (Figure 6): for example, its progenitor may be aligned with the major axis of the cluster *ab initio*, and its orientation will not have changed since its formation. Likewise, numerical simulations (Villumsen, 1982, 1983) indicate that a head-on collision of galaxies produces a final object which is highly prolate in shape, and whose major axis is oriented along the initial separation vector of the two merging objects.

Thus, cD galaxies are predicted to have prolate shapes, in agreement with determinations of the distribution of ellipticities observed for the brightest galaxies in clusters by Ryden, Lauer, & Postman (1993). The brightest galaxies tend to be flatter and more prolate than any other member of the cluster. Moreover, numerical simulations make evident that a large number of non-negligible size clumps fall into the cluster from the direction dictated by the host filaments. Before the cluster reaches the virialization, substructures should dominate the internal dynamics, *e.g.*, perturbing gas density detectable by X-ray. These could be the substructure of clusters detected by Kolokotronis et al., (2001), and Plionis & Basilakos (2002). In fact, for the anisotropic merger process to form clusters and filaments, results natural for substructure to get aligned with the cluster since the forming material clumps along that direction (Schuecker et al., 2001).

From the observational viewpoint, one could perform a temperature map of clusters, addressed to detect any anisotropy pattern of heating, which would be indicative of an on-going merger between a non-negligible size substructure and the main cluster. When two peaks are detected, and confirmed their "isophotal" (bridge) connection, a preferential direction for galactic merger should be detected. Such an idea has been pursued by Henriksen et al., (2000) for the case of A3266, who measured a velocity dispersion for the bridge of $\sim$ 1300 km/s compared with $\sim$ 1000 km/s for the surrounding cluster (very similar to Hill et al., 1988). Additionally, they used the optical and the X-ray observations to shown that a loose group of 30 galaxies has penetrated the main cluster from the direction suggested by the bridge, decoupled from their original intergroup medium and passing through a strong shock front. The substructure detected in A754 by Godlowski et al., (1998); and Godlowski and Ostrowski(1999), showed a strong tendency for non-random alignment and confirm the presence of subclustering within this cluster.

Together with preserving the orientation of the peak after the anisotropic accretion of material, there are other equally important alignment effects that enter into play: the cluster-cluster alignment, and the anisotropy of the cluster-galaxy correlation function. The former of these is usually adopted as evidence of the formation of filamentary structures, whereas the latter – which is a measure of the anisotropic clustering of galaxies located preferentially along the cluster's main axes – is a consequence of the clustering of clusters. These three types of alignment suggest a large-scale coherence of the density field where the orientation of peaks on the scale of clusters – and probably cD and other galaxies – has not changed considerably. Bond (1986, 1987a,b,c, 1996) and Pogosyan et al., (1996) have suggested that a primordial anisotropy of the density peaks would be re-enhanced by the non-linear evolution of the density field.

Fong, Stevenson and Shanks (1990) found some statistical evidence for three types of alignment :

(a) between neighboring galaxies in a cluster;

(b) between the major axes of galaxies and the cluster's major-axis;

(c) between the position angle of a galaxy and the radial vector to the cluster center.

The last two types of alignment were found to be of modest significance.

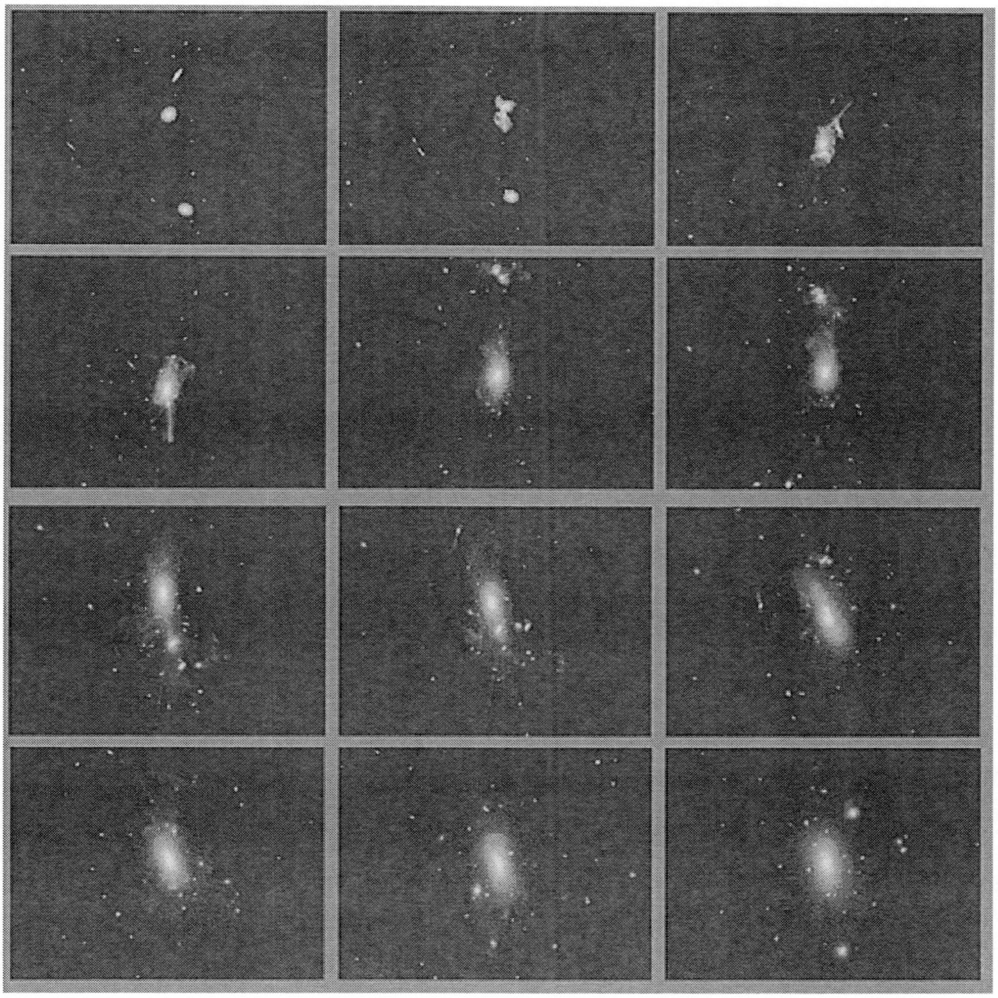

Figure 6. cD galaxy formation by anisotropic mergers. The cDs orientation preserves the direction of the initial matter flow. From Dubinski, 1998.

6.4. Environmental Influence on Galaxy Orientations

We can summarize the existence of at least five factors which can influence the orientation of galaxies in clusters :

(a) Strong interactions in high density environments with neighbors, which pass closer than the mean separation of galaxies,

(b) Tidal interactions with more distant galaxies,

(c) Tidal interactions produced by the smooth mass distribution in the cluster,

(d) Dynamical friction,

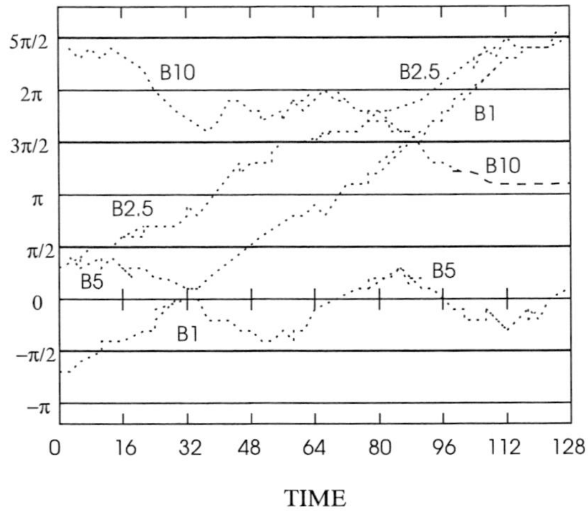

Figure 7. Bar major axis angle (ordinate) as a function of time. The bar oscillated in experiments B5 and B10, but it continues rotating in experiments B1 and B2.5. Horizontal lines corresponds to parallel alignment of the bar with the radial vector. Short vertical lines indicates that bar is perpendicular to the radial vector. The rotating bar dwell briefly near the perpendicular orientation.

(e) The character of the galaxy orbit.

Concerning (a), this type of interaction can only affect galaxies located in the core of clusters where the frequency of collisions is higher than that in the envelope. The study of Farouki & Shapiro (1981), on the influence on galaxies of rapid tidal encounters, showed that they are not relevant in changing the orientations of galaxies. Interactions of type (b) may not be significant because the galaxies are distributed approximately randomly in space, so that the torques produced by them should not agree in direction. However, the influence on galaxies of a strong tidal field produced by the core of the cluster could be important – as suggested by the results of Miller & Smith (1982, 1988).

The radial character of the orbit of a galaxy is also of crucial importance in the study of tidal interactions. If the orbit were not radial the dynamics of the galaxy in the presence of a time-varying torque will be more complex than the situation studied by Miller & Smith (1982, 1988) and Farouki & Shapiro (1981). The former of these studies have considered the effects on the internal dynamics of rotating galaxies produced by the tidal field of their host clusters. They used 79840 particles in order to model a bar-like galaxy with nonzero-angular momentum, to which an initial kinetic energy (T) associated with the mean (fluid) motions and a gravitational energy (W) are assigned. The ratio of these two quantities, $\tau \sim T/\mid W \mid = v^2/2\sigma^2$, measures the growth of the rotational velocity relative to the velocity dispersion of particles.

Miller et al., (1982,1988) adopted a quadratic potential, $\Phi = (k_l x^2 + k_t y^2 + k_t z^2)$ – where the degree of anisotropy of the tidal field is determined by the difference $k = \mid k_t - k_l \mid$. The response of a galaxy to this field is the sum of the responses of each particle. For strong torques such as those produced by the cluster core, the tidal force on the particles is greater than the gravitational force responsible for the internal galaxy dynamics. The galaxy could therefore be deformed or even disrupted. Miller & Smith (1982, 1988) found that after a Hubble time an initial ratio $\tau = 0.5$ decayed to $\tau < 0.2$ – which indicates a considerable reduction in the rotation of the galaxy. At the same time, the shape the galaxy was only slightly deformed. They found that the coupling of the orbits of particles to the field is highly coherent, and the final effect of this is to replace the bar rotation pattern by oscillations that are damped out so as to leave the bar pointing toward the cluster center (Figure 2). For weak tidal fields the disruption is weaker and the bar performs oscillations around radial alignment. Although this kind of interaction can, in principle, account for the radial alignments in clusters, these results greatly depend on the value of the parameter $\tau = T / \mid W \mid$. From all the galaxies in a cluster one may not have many objects with the *ad hoc* value of $\tau \approx 0.5$ which is necessary to show radial alignment. Further, parallel alignments are not accounted by this mechanism. In these studies, stiffness of the bar was questioned by Farouki & Shapiro (1981) who performed N-body simulations to study the morphological changes of disk galaxies in dense clusters due to rapid tidal encounters. They followed the passage of a spiral galaxy through the core of a cD cluster, finding that this kind of interaction produces morphological distortions in the outer regions, but leaves its central region unperturbed. Besides, the position angle of the galaxies and their spin-vectors are not considerably changed. They pointed out that these results would indicate that orientations could be preserved fossil quantities.

One of the observational methods to detect the influence of the cluster center potential on galaxies is by detecting an isophotal twisting of galaxies. There have been several photometric studies to search for correlations in the properties of elliptical galaxies – such as their ellipticities or the PA of their major axes with their positions in the host cluster. Ciotti & Dutta (1994) have studied both analytically using numerical simulations, to examine the influence of a cluster tidal field on the morphology and orientation of elliptical galaxies – moving either in radial or circular orbits. They studied two effects: the change in ellipticity of the isophotal contours of the galaxies as a function of the distance to the cluster center, and a possible isophotal alignment of a prolate galaxy with the radial vector to the cluster center. The cluster is treated as an externally-imposed, homogeneous, spherically symmetric stationary field, whereas the galaxy – considered spherical or prolate – is evolved for many crossing times. They found that topological changes in the cluster tidal field near its core radius cause galaxies on radial orbits crossing this region to see non-adiabatic changes in the potential – *i.e.*, the tidal field is not greatly modified, and the alignment times for prolate galaxies are much longer than their dynamical timescales. Their main results are that the morphological changes produced by the cluster tidal field on spherical models are not detectable up to many effective radii. However, a substantial major axis alignment with the cluster radial direction is reached in a few galactic dynamical times for prolate galaxies on circular orbits outside the cluster core radius. Such an alignment is produced by the torque induced by the tidal field on the galaxy inertia tensor, and also by the conversion into disordered stellar motion of the ordered rotational energy acquired by the galaxy. Thus, Ciotti & Dutta predict population segregation as a function of distance of the galaxy from the cluster center, with galaxies outside the cluster core showing a greater tendency to have their major axes aligned in a few internal dynamical times with the direction to the cluster center.

Another interesting idea to explain radial and parallel alignment of galaxies has been put forward by Bhattacharjee, Ghosh & Cowsik (1989) and Ghosh, Bhattacharjee & Cowsik (1989), who attempted to explain alignment of galaxies as a gravitational thermodynamical system. They concluded – excluding morphological changes, radial, or circular orbits – that alignments can arise as the result of a gravitational phase transition, quite similar to those present for instance in nemetic liquid crystals.

As we have seen, several results show the possibility of alignments between neighboring galaxies, or between the major axes of galaxies and the radial vector to the center of the parent cluster. However, there are few guesses about the origin of alignments parallel to the cluster major axis. In the next section, we will explain in detail the possible mechanism for this kind of alignment, which seems to be closely related with the origin of large-scale filaments.

7. Conclusion

In this chapter, we have compiled the observational evidence for a wide gamut of alignments of cosmic structures on small and large scales, which involve either their major axes, their angular momentum vectors or the direction to the closest neighbors. Whereas tidal interaction has been frequently invoked to generate such alignment effects, several studies seem to confirm that small-scale tidal torques cannot account for all the types of observed alignments (e.g., González & Teodoro, 2010). Furthermore, the successful theory of cD galaxy formation by anisotropic mergers of smaller gas-rich protogalactic units can account for the observed alignment of a cD galaxy with its host cluster major axis, and suggest that the material necessary to form it moves in the direction dictated by the formation of large-scale filaments. Major support for this theory is given by the primordial alignment effects detected by Bond et al. (1996) using the peak-patches theory. See also González (1995).

The existing evidence of a correlation of the orientation of the innermost regions ($2h^{-1}$ Mpc) of BCGs with the surrounding distribution of matter – over distances of up to $\sim 15h^{-1}$ Mpc – indicates that information about the distribution of neighboring clusters has been communicated to the cluster centers. It is possible that some of this information is still preserved by some fraction of the cluster's galaxies. In the non-linear evolution of the cluster, galaxies are affected by dynamical friction and tidal interactions with the cluster potential, which can change their initial orientation and morphology. However, within the context of the observational evidence for alignment of galaxies in clusters and superclusters, it has been proposed that the first galaxies to form are born aligned with the main plane or with the direction of their host structures. Subsequently, as the parent structure continues to collapse, tidal interactions become more important, gradually erasing any memory of the initial conditions. Some oversimplified estimates have suggested that the angular momentum vector makes the galaxy orientation stable for longer than a Hubble time. For low or null angular momentum systems orientations are unstable but are still on scales of the order of a Hubble time. Although this picture – of small-scale alignments – is perhaps naive, it nonetheless provides some points worthy of further consideration. However, for gravitational lensing purposes, the intrinsic alignment of cosmic structures –at least at the galaxy level– is an unwished ingredient which should be much better understood. In a second part of this subject, in this volume or another yet to come, we will develop the theoretical frame of work to analyze the primordial density field and give some evidence in favor that several but not all the alignment effects can survive the high nonlinear evolution.

Acknowledgments

This work was supported by CONACYT, SNI, and PROMEP (México). The authors would also like to thank NovaScience for publishing this manuscript.

References

[1] Aarseth, S. J. & Binney, J.1978. *MNRAS*, **185**, 227.

[2] Abbe, C.1875.*American Journal of Science and Arts*, **9** (109), 3th ser. 42

[3] Adams, M.T., Strom, K.M., & Strom, S.E.1980. *ApJ*, **238**, 445.

[4] Adler, R.J.1981. 'The Geometry of Random Fields,' (Wiley, Chichester)

[5] Allen, S.W., Fabian, A.C., Edge, A.C., Bohringer, H. & White, D.A.1995. Cooling Flows, Central Galaxy-Cluster Alignmnets, X-ray absortion and Dust, *MNRAS*, **275**, 741.

[6] Argyres, P.C., Groth, E. J., Peebles, P.J.E.,& Struble, M.F. 1986. *AJ*, **91**, 471.

[7] Arp, H. 1968. *PASP*, **80**, 129.

[8] Arp, H.1990. *Galaxy Alignments, Journal of Astrophysics and Astronomy,* 11, 411.

[9] Bacon, D.J., Refregier, A.R., & Ellis, R.S. 2000. *MNRAS*, **318**, 625.

[10] Bahcall, N.A., 1987. In 'Observational Cosmology', Proceedings of the IAU Symposum, Beijing. (Reidel, Dordrecht).

[11] Bardeen, J.M., Kaiser, N., Bond, N., & Szalay, R. 1986. *ApJ*, **304**, 148.

[12] Barnes, J. E. 1992. *ApJ*, **393**, 484.

[13] Barnes, J. A., & Efstathiou, G.1987. *ApJ*, **319**, 575.

[14] Bartelman, N., & Schneider, P. 1999. astro-ph/9912508.

[15] Batuski, D.J., & Burns, J.O. 1985. A Possible 300 Mpc Filament of Clusters of Galaxies in Perseus-Pegasus. *ApJ*, **299**, 5.

[16] Bernardeau, F., van Waerbeke, L. & Mellier, Y. 1997. *AAp*, **322**, 1.

[17] Bhattacharjee, P., Ghosh, P., & Cowsik, R. 1989. *ApJ*, **339**, 638.

[18] Binggeli, B. 1982. *AAp*, **107**, 338.

[19] Binggeli, B.1999. In Ringber workshop, "The Radio Galaxy Messier 87," Ed. H.J. Roser & K. Meisnhermer (Berlin, Springer), 9.

[20] Binggeli, B., Tammann, & Sandage, A. 1987. *AJ*, **94**, 251.

[21] Binney, J. 1974. *MNRAS*, **168**, 73.

[22] Binney, J., & Tremaine, S. 1987. 'Galactic Dynamics,' (Princeton University Press.)

[23] Binney, J. & Silk, J. 1979. *MNRAS*, **188**, 273.

[24] Blumenthal, G.R., Faber, S.M., Primack, J.R., & Rees, M.J. 1984. *Nature*, **311**, 517.

[25] Bond, J.R. 1986. In 'Galaxy Distances and Deviations from the Hubble Flow,' Eds., B.F. Madore and R.B. Tully, (Reidel, Dordrecht).

[26] Bond, J.R. 1987a. In 'Nearly Normal Galaxies,' Ed., S. Faber, (Springer Verlag, New York).

[27] Bond, J.R. 1987b. In 'The early Universe,', Ed., W.G. Unruh, (Reidel, Dordrecht).

[28] Bond, J.R. 1987c. In 'Cosmology and Particle Physics,', Ed., I. Hinchcliffe, (World Scientific, Singapore)

[29] Bond, J.R. & Myers, S.T. 1996. The Peak-patch Picture of Cosmic Catalogs. III. , *ApJS*, **103**, 63.

[30] Bond, J.R., Kofman, L., & Pogosyan, D. 1996. *Nature*, **380**, 603

[31] Brown, F.G. 1939. *MNRAS*, **99**, 534.

[32] Brown, F.G. 1964. *MNRAS*, **127**, 517.

[33] Brown, F.G. 1968. *MNRAS*, **138**, 527.

[34] Brown, M.L., Taylor, A.N., Hambly, N.C., & Dye, S. 2002. *MNRAS*, **333**, 501.

[35] Cabanela, J.E., & Aldering, G. 1994a. *Bull. AAS*, **184**, 6208.

[36] Cabanela, J.E., & Aldering, G. 1994b. *Bull. AAS*, **185**, 5507.

[37] Cabanela, J.E., & Aldering, G. 1998. Galaxy Alignments in the Pisces-Perseus Supercluster Revisited. *ApJ*, **116**, 1094.

[38] Caimmi, R. 1989. *AAp*, **223**, 29.

[39] Carlberg, R.G. 1993. *AAp Lett.*, **411**, L9.

[40] Catelan, P. & Porciani, C. 2001. Correlations of Cosmic Tidal Fields, **323**, 713.

[41] Catelan, P. & Theuns, T. 1996. *MNRAS*, **282**, 436.

[42] Catelan, P., Kamionkowski, M., & Blandford, R.D. 2001. Intrinsic and Extrinsic Galaxy Alignment, *MNRAS*, **320**, L7.

[43] Carter, D. & Metcalfe, N. 1980. *MNRAS*, **191**, 325.

[44] Cerne, J. & Peterson, J.B. 1990. *AJ*, **100**, 1761.

[45] Chambers, S.W., Mellot, A.L., & Miller, C.J. 2000. Einstein Cluster Alignments Revisited, *ApJ*, **544**, 104.

[46] Chapman, G.N., Geller, M.J., & Huchra, J.P. 1987. *AJ*, **94**, 571.

[47] Chapman, G.N., Geller, M.J., & Huchra, J.P. 1988. *AJ*, **95**, 999.

[48] Ciotti, L., & Dutta, S.N. 1994. *MNRAS*, **270**, 390.

[49] Cooray, A., & Hu, W. 2000. astro-ph/0004151

[50] Coutts, A. 1996. The Scale and Dispersion of Galactic Alignments. *MNRAS*, **278**, 87.

[51] Crittenden, R.G., Nataraja, P., Pen, U., & Theuns, T. 2001. *ApJ*, **559**, 552.

[52] Crittenden, R.G., Nataraja, P., Pen, U. & Theuns, T. 2002. *ApJ*, **568**, 20.

[53] Croft, R.A.C., & Metzler, C.A. 2000. *ApJ*, **545**, 561.

[54] Cross, L.L., & David, L.P. 1995. The Shapes of Clusters of Galaxies Within Superclusters, *Bull. AAS*, **187**, 54.03.

[55] de Theije, P.A.M., van Kampen, E., & Slijkhuis, R. 1997. preprint

[56] Dekel, A., & Aarseth, S.1984. *ApJ*, **238**, 1.

[57] Dekel, A. 1985. *ApJ*, **298**, 461.

[58] Dekel, A. 1988. In 'Large-scale Motions in the Universe,' Eds., Vera C. Rubin and G. V. C. Coyne, (Princeton University Press)

[59] Dekel, A., West, M., & Aarseth, S. 1984. *ApJ*, **279**, 1.

[60] Djorgovski, S. 1983a. *Bull. AAS*, **15**, 656.

[61] Djorgovski, S. 1983b. *ApJ*, **274**, L7.

[62] Djorgovski, S. 1987. In 'Nearly Normal Galaxies,' Ed., S. Faber., (Springer-Verlag, New York), 226.

[63] Doroshkevich, A.G. 1970. *Astrofizika*, **6**, 581.

[64] Doroshkevich, A.G. 1973a. *ApJ* **14**, L11.

[65] Doroshkevich, A.G. 1973b. *Soviet Astr.*, **16**, 986.

[66] Doroshkevich, A.G. & S.F. Shandarin, 1978. *Soviet Astr.*, **22**, 6.

[67] Doroshkevich, A.G., Shandarin, S.F., & Saar, E. 1978. *MNRAS*, **184**, 643.

[68] Dubinski, J. 1992. *ApJ*, **401**, 441.

[69] Dubinski, J. 1998. The Origin of the Brightest Cluster Galaxies. *ApJ*, **502**, 141.

[70] Efstathiou, G.P.,& Jones, B.J.T. 1979. *MNRAS*, **186**, 133.

[71] Efstathiou, G.P., Davis, M., Frenk, C.S., & White, S.D.M. 1985. *ApJ*, **292**, 371.

[72] Efstathiou, G.P. 1990. In 'The Physics of the Early Universe,' Eds., J.A. Peacock, A.F. Heavens, & A.T. Davies (SUSSP publications).

[73] Einasto, J., Gramann, M., Saar, E., & Tago, E. 1993. *MNRAS*, **260**, 705.

[74] Einasto, J., Joeveer, M., & Saar, E. 1980. *MNRAS*, **193**, 353.

[75] Elmegreen, B.G., & Elmegreen, D.M. 1978. *ApJ*, **220**, 1051.

[76] Evrard, A.E.1989. *ApJ*, **341**, 26.

[77] Evrard, A.E., Silk, J., & Szalay, A.S. 1990. *ApJ*, **365**, 13.

[78] Faber, S.M. 1982. In 'Astrophysical Cosmology,' Ed., Bruck et al., (Ponti. Acad. Sci. Vatican City), 191

[79] Faltenbacher, A., Li, C., Mao,S., van den Bosch, F.C.,Yang, X., Jing, Y.P.,Pasquali, A., & Mo, H.J. 2007. *ApJ*, **662**, L71.

[80] Faltenbacher, A., Kerscher, M., Gottloeber, S.,& Mueller, M. 2002. *AAp*, **395**, 1.

[81] Farouki, R. & Shapiro, S.L. 1981. *ApJ*, **243**, 32.

[82] Fitchett, M.J. 1988. In 'The Minnesota Lectures on Clusters of Galaxies and Large-scale Structure,' Eds., J.M. Dickey, (Brigham Young Univ. Press), 143

[83] Flin, P. 1987. *MNRAS*, **228**, 941.

[84] Flin, P. 1988a. *MNRAS*, **235**, 857.

[85] Flin, P. 1988b. In 'Large-scale Structure of the Universe,' Ed., J Audouze, (IAU), 239

[86] Flin, P. 1993. *AJ*, **105**, 473.

[87] Flin, P. & Godlowski, W. 1984. In 'Clusters and Groups of Galaxies,' Eds., F. Mardirossian, G. Giuricin, and M. Mezzeti (Reidel, Trieste).

[88] Flin, P. & Godlowski, W. 1986. *MNRAS*, **222**, 525.

[89] Flint, K.P., & Impey, C.D. 1996. Large-scale Structure in the Direction of the Virgo Cluster. *AJ*, **112**, 865.

[90] Fong, R., Stevenson, P.R., & Shanks, T. 1990. *MNRAS*, **242**, 146.

[91] Fontanelli, P. 1984. *AAp*, **138**, 85.

[92] Frenk, C.S. 1990. In 'The Birth and Early Evolution of our Universe,' Ed., B. Skagertsten, (Graftavallen, Sweden).

[93] Frenk, C.S., White, S.D.M., Davis, M.,& Efstathiou, G. 1985. *Nature*, **317**, 595

[94] Frenk, C.S., White, S.D.M., Davis, M.,& Efstathiou, G. 1988. *ApJ*, **227**, 507.

[95] Fuller, T.M., West, M.J.,& Bridges, T.J. 1999. Major Axis Alignments of Poor Cluster Dominant Galaxies. *ApJ*, **519**, 22.

[96] Gainullina, R. & Roshjakova, T.V. 1967. *Trudy. Inst. Ap. Akad. Nauk. Kaz.* **5**, 237

[97] Garijo, A., Athanassoula, E., & García-Gómez, C. 1997. *AAp*, **327**, 930.

[98] Garrido, J.L., Battaner, E.S., Sánchez-Saavedra, M.L., & Florido, E. 1993. *AAp*, **271**, 84.

[99] Ghosh, P., Bhattacharjee, P., & Cowsik, R. 1989. *AJ* 345, 647.

[100] González, A. 1995. D. Phil. thesis, Cosmological Alignment Effects. University of Sussex, England.

[101] González, A. 1995a. In 'Fresh Views of Elliptical Galaxies,' Eds., A. Buzzoni, A.Renzini, & A.Serrano, ASP Conf. Ser. **86**, 73

[102] González, A. & Teodoro, L.F.A. 2010. *MNRAS*, **404**, L11

[103] Gregory, C. 1924. *MNRAS*, **84**, 456.

[104] Gregory, S.A., & Thompson, L.A. 1978. *Nature*, **274**, 450.

[105] Gregory, S.A., Thompson, L.A., & Tifft, W. 1981. *ApJ*, **243**, 411.

[106] Godlowski, W., & Baier, F. 1997. Galaxy Orientations in Some Abell Clusters, *Proceedings of the 186 IAU Symposium, 410*

[107] Godlowski, W., Baier, F.W., & MacGillivray, H.T. 1998. Substructures and Galaxy Orientations in Clusters.I. The Cluster Abell 754. *AAp*, **339**, 709.

[108] Godlowski, W., & Ostrowsi, M. 1999. Investigation of Galactic Alignment in Local Supercluster galaxy clusters. *MNRAS*, **303**, 50.

[109] Hatton, S., & Ninin, S. 2001. Angular Momentum alignment of Dark Matter Haloes. *MNRAS*, **322**, 576.

[110] Hawley, D.L., & Peebles, P.J.E. 1975. *AJ*, **80**, 477.

[111] Heavens, A., & Peacock, J. 1988. *MNRAS*, **232**, 339.

[112] Heavens, A., Refregier, A., & Heymans, C.2000. Intrinsic Correlations of Galaxy Shapes: Implications for Weak Lensing, *MNRAS*, **319**, 649.

[113] Heymans, C., & Heavens, A. 2002. Weak Gravitational Lensing: Reducing the Contamination by Intrinsic Alignments, astro-ph/0208220.

[114] Helou, G. 1984. *ApJ*, **284**, 471.

[115] Henriksen, M., Donnelly, R., Hank, R., & Davis, S. 2000. Reconstructing the Merger History of the A3266 Galaxy Cluster. *ApJ*, **529**, 692.

[116] Hill, J.M., Hintzen, P., Oegerle, W.R., Romanishim, W., Lesser, W.P., Eisenhamer, D.J., & Batuski, D.J. 1988. *ApJ*, **322**, L23.

[117] Hoekstra, H., Yee, H.,& Gladders, M. 2001. *ApJ*, **3558**, L11.

[118] Hoekstra, H., Yee, H., & Gladders, M. 2002. astro-ph/0205412

[119] Hu, W., & Tegmark, M.1999. *ApJ*, **514**, L65.

[120] Hui, L., & Zhang, J. 2002. astro-ph/0205512.

[121] Iye, M., & Sugai, H. 1991. *ApJ*, **374**, 112.

[122] Jaaniste, J., & Saar, E. 1977. Tartu Obs. preprint A-2.

[123] Jing, Y.P. 2002. Intrinsic Correlation of Halo Ellipticities and its Implications for Large-scale Weak Lensing Survey, astro-ph/0206098

[124] Jing, Y.P., Plionis, M., & Valdarnini, R. 1992. *ApJ*, **389**, 499.

[125] Joeveer, M., Einasto, J., & Tago, E. 1978. *MNRAS*, **185**, 357.

[126] Kaiser, N. 1998. *ApJ*, **498**, 26.

[127] Kapranidis, S., & Sullivan, W. 1983. *AAp*, **118**, 33.

[128] Kashikawa, N. & Okamura, S. 1992. *PASJ*, **44**, 493.

[129] Kim, R.S.J. 2001. (SDSS collaboration), astro-ph/0110383

[130] Kitzbichler, M.G., & W. Saurer, W. 2003. Detection of Non-random Galaxy Orientations in the X-ray Subclusters of the Coma Cluster, astro-ph/0303016

[131] Kofman, L., Pagosyan, D., Shandarin, S.F., & Melott, A. 1992. *ApJ*, **393**, 437.

[132] Kolokotronis,V., Basilakos, S., Plionis, M., & Georgantopoulos, I. 2001. Searching for Cluster Substructure Using APM and ROSAT data. *MNRAS*, **320**, 49.

[133] Knox, S. 1908. *MNRAS*, **69**, 72.

[134] Lambas, D.G., Groth, E.J., & Peebles, P.J.E. 1988a. *AJ*, **95**, 975.

[135] Lambas, D.G., Groth, E.J., & Peebles, P.J.E. 1988b. *AJ*, **95**, 996.

[136] Lambas, D.G., Nicotra, M., Muriel, H., & Ruiz, L. 1990. *AJ*, **100**, 1006.

[137] Lauer, T.R. 1988. *ApJ*, **325**, 49.

[138] Lauer, T.R. 1990. In 'Dynamics and Interactions of Galaxies,' Ed., R. Wielen, (Springer, Heidelberg), 406

[139] Lee, J., & Pen, U.-L. 2000a. Cosmic Shear from Galaxy Spins, *ApJ Sup.*, **532**, L55.

[140] Lee, J. & Pen, U.-L. 2000b. astrop-ph/0008135

[141] Laubscher, B.E., & Gregory, S.A. 1993. *Bull. AAS*, **913**, 25.

[142] Little, B.,Weinberg, D.H.,& Park, C. 1991. *MNRAS*, **253**, 295.

[143] MacGillivray, H.T., & Dood, R.J. 1985. *AAp*, **145**, 269.

[144] MacGillivray, H.T., Dood, R.J., Lightfoot, J.F., Trew, A.S., & McNally, B.V. 1980. Monte-Carlo Simulation of the Orientations of Galaxies in Clusters, *APSS*, **70**, 385.

[145] MacGillivray, H.T., Dood, R.J., McNally, B.V., & Corwin, H.G. 1982. *MNRAS*, **198**, 605.

[146] Mackey, J., White, M., & Kamionkowski, M. 2001. Theoretical Estimates of Intrinsic Galaxy Alignmens, astro-ph/0109063

[147] Malumuth, E.M., Smith, E.P., & Kriss, G.A. 1989. *Bull. AAS*, **21**, 1168.

[148] Malumuth, E.M. 1992. *ApJ*, **386**, 420.

[149] Mandzhos, A.V., Tel'nyuk-Adamchuk, V.V., & Gregul, A.Y. 1985. *Soviet Astr.* **11**, 206.

[150] Martin, D.R., Nichol, R.C., Collins, C.A. Lumsden, S.A., & Guzzo, L. 1995. Cluster Alignments in the Edimburgh-Milano Cluster Redshift Survey. *MNRAS*, **274**, 623.

[151] McMillan, S.L., Kowalski, M.P., & Ulmer, P. 1989. *ApJ Sup*, **70**, 723.

[152] Merritt, D. 1983. *ApJ*, **264**, 24.

[153] Merritt, D. 1984. *ApJ*, **276**, 26.

[154] Merritt, D. 1985. *ApJ*, **289**, 18.

[155] Meyermann, B. 1923. *Astron. Nachr.* 219, 133.

[156] Mellot, A.L., & Shandarin, S.F. 1989. *ApJ*, **343**, 26.

[157] Miller, R.H., & Smith, B.F. 1982.*ApJ*, **253**, 58.

[158] Miller, R.H., & Smith, B.F. 1988. In 'The few body problem,' IAU Colloquium 96, Ed., M.J. Valtoner (Kluwer, Boston) 353

[159] Millington, S.J.C., & Peach, J.V. 1990. The Rich Cluster of Galaxies A348. *MNRAS*, **242**, 112.

[160] Miralles, J.M., Erben, T., Hammerle, H., Schneider, P., Fosbury, R.A.E., Freudling, W., Pirzkal, N., Jain, B., & White, S.D.M. 2002. A Conspicuous tangential alignemnt of galaxies in a STIS parallel shear survey field: A new dark-lens candidate?, astro-ph/0202122

[161] Muriel, H., & Lambas, D.G. 1989. *ApJ*, **98**, 1995.

[162] Muriel, H., & Lambas, D.G. 1992. *AJ*, **103**, 393.

[163] Munshi, D. & Jain, B. 2000. *MNRAS*, **318**, 109.

[164] Nilson, P. 1973. Uppsala General Catalogue of Galaxies (Univers Uppsala, Uppsala.)

[165] Noonan, T. 1980. *ApJ*, **238**, 793.

[166] Novikov,D., Melott, A.L., Wilhite, B.C., Kaufman, M., Burns, J., Miller, O., & Batuski, D.J. 1999. Cluster Winds Blow Along Superclusters Axes. *MNRAS*, **304**, L5.

[167] Olson, D.W. 1980. *ApJ*, **236**, 335.

[168] Onoura, L.I., & Thomas, P.A. 2000. The alignment of Clusters Using Large-scale Simulations. *MNRAS*, **319**, 614.

[169] Oort, J.H. 1984. In 'Clusters and Groups of Galaxies,' Eds., F. Mardirossian, G. Giuricin, & M. Mezzeti, (Reidel, Trieste)

[170] Oosterloo, T.A. 1988. D. Phil. thesis, University of Groningen

[171] Ostriker, J.P., & Hausman, M.A. 1977. *ApJ*, **217**, L125.

[172] Pauls, J.L., & Mellot, A.L. 1995. *MNRAS*, **274**, 99.

[173] Padmanabhan, T. 1993. 'Structure Formation in the Universe,' (Cambridge University Press)

[174] Peacock, J A.,& Heavens, A.F. 1985. *MNRAS*, **217**, 805.

[175] Peacock, J A. & West, M.J. 1992. *MNRAS*, **259**, 494.

[176] Peebles, P.J.E. 1969. *ApJ*, **155**, 393.

[177] Peebles, P.J.E. 1980. 'The Large-scale Structure of the Universe,' (Princeton University Press)

[178] Pereira, M.J., & Khun, J.R. 2005. *ApJ*, **627**, L21.

[179] Pereira, M.J., Bryan, G.L., & Gill, S.P. 2008. *ApJ*, **672**, 825.

[180] Pen, U.-L., Lee, J., & Seljak, U. 2000, Tentative Detection of Galaxy Spin Correlations in the Tully Catalog. *ApJ*, **543**, L107.

[181] Plionis, M., 1989. D. Phil thesis, University of Sussex, England.

[182] Plionis, M.,1994. Position angles and alignments of clusters of galaxies. *ApJS*, **95**, 401.

[183] Plionis, M.,& Basilakos, S. 2002. The Cluster Substructure-Alignment Connection. *MNRAS*, **329**, L47.

[184] Plionis, M., Maurogordato, S., Benoist, C., Ferrari, C.,& Basilakos, S. 2003. Galaxy Alignments as a Probe of the Dynamical State of Clusters, astro-ph/0305215

[185] Plionis, M., Valdarnani, R., & Jing, Y. 1992. *ApJ*, **398**, 12.

[186] Pogosyan, D., Bond, J.R., Kofman, L., & Wadsley, L. 1996. The Cosmic Web and Filaments in Cluster Patches, American, Astronomical Society Meeting, 189, 13.03

[187] Porter, A.C., Schneider, D.P., & Hoessel, J.G. 1991. *AJ*, **101**, 561.

[188] Quinn, T., & Binney, J.1992. *MNRAS*, **255**, 729.

[189] Read, J.I., Goerdt, T., Moore, B., Pontzen, A.P., Stadel, J. & Lake, G. 2006. *MNRAS*, **373**, 1451.

[190] Reinhardt, M. 1971. *APSS*, **10**, 363.

[191] Reinhardt, M., & Roberts, M.S. 1972. *Astrophys. Lett.* **12**, 201

[192] Rhee, G. & Roos, N. 1990. *MNRAS*, **243**, 629.

[193] Rhee, G., van Haarlem, M., & Katgert, P.1992. *AJ*, **103**, 1721.

[194] Rhee, G., & Katgert, P. 1987. *AAp*, **183**, 217.

[195] Rhodes, J., Refregier, A., & Groth, E.J. 2001. *ApJ*, **552**, L85.

[196] Rood, H.J., Page,T., Kintner, E.C., & King, I.R. 1972. *ApJ*, **175**, 627.

[197] Rood, H.J., & Sastry, G.N. 1971. *PASP*, **80**, 313.

[198] Ruffini, R., Song, D.J., & Taraglio, S. 1990. *AAp*, **232**, 7.

[199] Ryden, B.S. 1988a. *ApJ*, **329**, 589.

[200] Ryden, B.S.1988. *ApJ*, **333**, 78.

[201] Ryden,B.S., Lauer, T.R., & Postman, M.1993. *ApJ*, **410**, 515.

[202] Salvador-Sole, E., & Solanes, J.M. 1993. *ApJ*, **417**, 427.

[203] Sandage, A. 1976. *ApJ*, **205**, 6.

[204] Sastry, G.N. 1968. *PASP*, **80**, 252.

[205] Schindler, S., Binggeli, B.,& Boheringer, H. 1999. *AAp*, **343**, 420.

[206] Schneider, P., van Waerbeke, L.,& Mellier, Y. 2002. astro-ph/0112441

[207] Schueker, P., Bohringer, H., Reiprich, T.H., & Feretti, L. 2001. *AAp*, 378, 408.

[208] Seldner, M., & Peebles, P.J.E. 1977. *ApJ*, **215**, 703.

[209] Sharpley, J.M., Ellis, R.S., & Gray, P.M. 1988. *MNRAS*, **231**, 479.

[210] Sharp, N., Lin, D.N., & White, S.D.M. 1979. *MNRAS*, **187**, 287.

[211] Silk, J. & Evrard, A. 1988. *Nature*, **335**, 766

[212] Sofue, Y. 1992. *PASP*, **44**, L1.

[213] Splinter, R.J., Melott, A.L., Linn, A.M., Buck, C., & Tinker, J. 1997. The Ellipticity and Orientation of Clusters of Galaxies in N-body Experiments. *ApJ*, **479**, 632.

[214] Stewart, G.W., & Sun Ji-Guang 1990. 'Matrix Perturbation Theory,' (Academic Press)

[215] Strom, K.M., & Strom, S.E. 1978. *AJ*, **83**, 73.

[216] Struble M.F., 1987. *ApJ*, **317**, 668.

[217] Struble M.F., 1990. *AJ*, **99**, 743.

[218] Struble M.F., & Peebles, P.J.E. 1985. *AJ*, **90**, 582.

[219] Struble M.F.,& Peebles, P.J.E. 1986. *AJ*, **91**, 1474.

[220] Sugai, H., & Iye, M. 1995. *MNRAS*, **276**, 327.

[221] Tago, E., Einasto, J., & Saar, E. 1984. *MNRAS*, **206**, 559.

[222] Thompson, L.A. 1976. *ApJ*, **209**, 22.

[223] Thompson, L.A. & Gregory, S.A. 1978. *ApJ*, **220**, 809.

[224] Thormen, G. 1997. *MNRAS*, **290**,411.

[225] Torlina, L., De Propris, R., & West, M.J. 2007. *ApJ*, **660**,L97.

[226] Trejo-Alonso, J. 2008. Batchellor thesis,University of Guanajuato. Mexico.

[227] Tremaine, S.D., & Richstone, D.O. 1977. *ApJ*, **212**, 311.

[228] Tremaine, S.D. 1990. In 'Dynamics and Interactions of Galaxies,' Ed., R. Wielen (Springer-Verlag, Berlin), 394

[229] Trevése, D., Cirimele, G., & Flin, P. 1992a.*AJ*, **104**, 935.

[230] Trevése, D., Flin, P., Miglori, L., Hickson, P., & Pitella, G. 1992b. *AApS*, **94**, 327.

[231] Tucker, G.S. & Peterson, J.B. 1988. *AJ*, **95**, 298.

[232] Tully, R.B. 1986. *ApJ*, **303**, 25.

[233] Ueda, H., Itoh, M., & Suto, Y. 1993. *ApJ*, **408**, 3.

[234] Ueda, H., Shimasaku, K., Suginohara, T., & Suto, Y. 1994. *PASJ*, **46**, 319.

[235] Ulmer, M.P., McMillan, S.L.W., & Kowalski, M.P. 1989. *ApJ*, **338**, 711.

[236] Van de Weygaert, R. 1997. In 'Cosmological parameters and evolution of the universe,' IAU Symposium, 183.

[237] Van den Bergh, S. 1983. *PASP*, **95**, 275.

[238] Van Haarlem, M. & Van der Weygaert, R. 1993. *ApJ*, **418**, 544.

[239] Van Kampen, G., & Rhee, G.F. 1990. *AAp*, **237**, 283.

[240] Van Waerbeke, L., et al., 2000. *AAp*, **358**, 30.

[241] Villumsen, J.V. 1982. *MNRAS*, **199**, 493.

[242] Villumsen, J.V. 1983. *MNRAS*, **204**, 219.

[243] Vogeley, M.S., Park, C., Geller, M.J., & Huchra, J.P.1992. *Ap.J.(Lett)*, **391**, 5

[244] Voglis, N., & Hiotelis, N. 1989. *AAp*, **218**, 1.

[245] Wesson, P.S. 1982. *Vistas in Astronomy*, **26**, 225

[246] Wesson, P.S. 1984. *AAp*, **138**, 253.

[247] Wesson, P.S. 1985. *AAp*, **151**, 105.

[248] West, M.J. 1989. *ApJ*, **347**, 610.

[249] West, M.J. 1992. In 'Morphological and Physical Classification of Galaxies,' Ed., G. Longo (Kluwer, Dordrecht)

[250] West, M.J. 1994. *MNRAS*, **268**, 79.

[251] West, M.J. & Blakeslee, A. 2000. The Principal Axis of the Virgo Cluster. *ApJ*, 542, L27.

[252] West, M.J., Dekel, A., & Oemler, A. 1989. *ApJ*, **336**, 46.

[253] West, M.J., Jones, C., & Forman, W. 1995. *ApJ*, **451**, L5.

[254] West, M.J., Villumsen, J., & Dekel, A. 1991. *ApJ*, **369**, 287.

[255] West, M.J., & Schombert, J.M. 1993. preprint.

[256] West, M.J., Weinberg, D.H., & Dekel, A. 1990. *ApJ*, **353**, 329.

[257] White, S.D. 1984. *ApJ*, **286**, 38.

[258] White, S.D.M., & Frenk, C.S. 1991. *ApJ*, **379**, 52.

[259] White, S.D.M., Frenk, C.S., Davis, M., & Efstathiou, G. 1987a. *ApJ*, **313**, 505.

[260] White, S.D.M., Davis, M., Efstathiou, G., & Frenk, C. 1987b, *Nature*, **330**, 451.

[261] Wilkerson, M.S., & Romanishin, W. 1981. *ApJ*, **243**, 427.

[262] Wyatt, S., & Brown, F. 1955. *AJ*, **60**, 415.

[263] Zel'dovich, Y.B. **1970**. *AAp*, **5**, 84.

[264] Zel'dovich, Y.B., Einasto, J., & Shandarin, S.F. 1982. *Nature*, **300**, 407.

In: Galaxies: Evolution, Morphology and Dynamics
Editors: A. Guerrera Sanchez et al., pp. 47-62
ISBN: 978-1-62081-185-6

Chapter 2

ON THE NATURE OF COSMOLOGICAL ALIGNMENT EFFECTS OF GALAXIES, GROUPS AND CLUSTERS. II. THEORY

***Alejandro González-Sánchez*[1], *César A. Caretta*[2], *A. Gutiérrez-Rodríguez*[1] *and P. Rivera*[1]**
[1] Unidad Académica de Física, Universidad Autónoma de Zacatecas, Zacatecas, México
[2] Departamento de Astronomía, Universidad de Guanajuato, Guanajuato, México

Abstract

We have already done an extensive discussion on the observational evidence for non-random orientations of major axes in neighboring galaxies and clusters within filaments and superclusters. The origin of such a coherence might be encoded into the nonlinear evolution of the density field through tidal interactions. But also, as we argue here, it could already be imprinted, back in time in the primordial structure of Gaussian density fields, as a consequence of a quantifiable cross-talk from small to large scales. It will be shown that if any coherent length exists, it would mainly be determined by the primordial tidal shear, especially from that produced by larger structures. In this picture, interclusters filaments of galaxies are enhanced by the non-linear evolution of the density field, with residual substructures keeping memory of their anisotropic origin. We finally discuss the possibility of a common, primordial origin for the wide gamut of alignment effects.

Keywords: Alignments, galaxies, clusters, cosmology

1. Introduction

Even when cosmological models of cosmic structure formation, such as HDM, CDM or even WarmDM have been ruled out, alignment effects have been shown to be a natural result since long ago. This is also the case for the currently favored concordance ΛCDM model where alignments also appear (e.g., Libeskind et al. 2005; Basilakos et al. 2006; Lee & Evrard, 2007). The first consequence of this is that these effects cannot be used as a discriminator between cosmologies, and the second one is that we need to understand their origin from first principles.

The origin remains a subject of debate between two main proposals, and some variations; (1) the role of anisotropic mergers and infall of matter along filaments, and (2) the importance of the stress tidal interaction. Concerning the first of these, Atlay et al. (2006) quantified the relative importance of the two effects, finding that for most of the halos the alignments are caused by the tidal field. However, for cluster-size halos are the alignment effects driven by anisotropic mergers and infall of matter along filaments. A variation of this is to consider that both proposals, in fact, are not antagonic as proposed by Bond et al. (hereafter BBKS, 1986). Anisotropic mergers and infall of matter are dictated by the *primordial* tidal field (Hashimoto et al. 2008; Jones et al. 2010). Thus, the first question to answer is: how primordial? In this chapter, we address this question. In the first instance, the common ingredient used by the conventional cosmological models is that structure formation started from the evolution of a Gaussian random density field which is generated by an inflation process.

Let us remark that the difference between these two scenarios, lies in the question whether the connection to filaments has a major contribution?; or, is not enough? If so, there must be a correlation between the properties of the density field at large and small scales. In the next section, we give an example of such correlations, and apply it to some primordial density fields used as the starting point to follow the cosmic structure formation. Then, we show some evidence that alignments can be already imprinted in the initial conditions as a consequence of the *cross-talk* effect at different scales.

2. The Origin of Alignment Effects

2.1. A Theoretical Scale of Correlations

A correlation between small and large-scale density fields naturally arises as a general characteristic of Gaussian random fields. The correlation between two real-valued random variables X and Y is defined in terms of their root mean square σ_X^2 and σ_Y^2 (Adler, 1981) as

$$\gamma_{XY} = \sigma_{XY}^2/(\sigma_X \sigma_Y) \; . \tag{1}$$

It can be shown that Gaussian variables measured on different filtering scales are not independent (BBKS 1986; Padmanabhan 1993). In order to exhibit this correlation between the small and large-scale density field we use the rms of the field, characterized by a power law spectrum $P(k)$

$$\sigma_{ab}^2 \equiv < \delta_a \delta_b > = \int P(k) W_G(kr_a) W_G(kr_b) d^3k \; , \tag{2}$$

where W_G is the smoothing window function and r_a and r_b are the filtering scales. We can explicitly write Eq. 2 in terms of the spectral index and a Gaussian window as

$$\sigma_{ab}^2 = 4\pi A \int k^{n+2} \exp\left(-\frac{1}{2}k^2 r_a^2\right) \exp\left(-\frac{1}{2}k^2 r_b^2\right) dk \; . \tag{3}$$

Adopting equal filtering scales in Eq. 3, we obtain the mean square overdensity. But using two different scales gives us a measure of the normalized correlation (Eq. 1) which results in

$$\gamma_{ab}(n=1) = \left[\frac{2r_a r_b}{(r_a^2 + r_b^2)}\right]^2 , \quad (4)$$

$$\gamma_{ab}(n=0) = \left[\frac{2r_a r_b}{(r_a^2 + r_b^2)}\right]^{3/2} , \quad (5)$$

$$\gamma_{ab}(n=-1) = \frac{2r_a r_b}{(r_a^2 + r_b^2)} , \quad (6)$$

$$\gamma_{ab}(n=-2) = \sqrt{\frac{2r_a r_b}{(r_a^2 + r_b^2)}} . \quad (7)$$

The correlation value reaches $\gamma_{ab} = 1$ only when the same scale of filtering is used, *i.e.*, the peak is correlated with itself. On the contrary, $\gamma_a \rightarrow 0$ indicates that the two scales are uncorrelated. We must realize and remark that evolution of the density field is not considered.

In order to express the correlation parameter as a function of the mass, we adopt an $\Omega = 1$, Einstein-de Sitter Universe, where the mass within a sphere of radius r is

$$M(r) = (2\pi)^{3/2} r^3 \bar{\rho} = 4.37 \times 10^{12} r^3 h^{-1} M_\odot . \quad (8)$$

Figure 1 shows that a considerable correlation (cross-talk) exists between cluster and supercluster scales, and is more prominent in density fields with flat spectra, $n \approx -1, -2$ and for high density peaks. By using numerical simulations, Little, Weinberg & Park (1991) detected that an initial correlation between different scales becomes stronger for high peaks, and further enhanced by the nonlinear coupling of Fourier modes, transferring more power from large to small scales. This means, first, that the formation of rich galaxy clusters, as well as their shape and orientation, can be strongly influenced by the large-scale density field and not only by its local properties. Second, these correlations can possibly be producing the galaxy bridges connecting clusters of galaxies detected first by Bahcall (1987).

2.2. Initial Conditions and Numerical Simulations

Bond (1986, 1987a, 1987b, 1987c), Gonzalez (1995), Bond & Mayers (1996) and Bond, Kofman & Pogosyan (1996) have shown that Gaussian cross-talk can produce strong anisotropies in the density field around cluster-scale peaks, which dictate the way in which the clusters are born aligned with their surroundings up to distances of tens of Mpc (*e.g.*, West et al. 1989; 1990; 1991; 1995). Thus, the filamentary web of the universe would be a consequence of the primordial distribution and spatial coherence just enhanced by the stress field in the medium. In order to show up the coherence in the field and how this becomes stronger during its evolution, by using the peak-patches theory developed by Bond & Myers (1996, and references therein), Bond et al. (1996) calculated the conditional multiple-correlation functions in Lagrangian space – *i.e.*, statistically average density and displacement fields subject to different constraints on the shear at multiple points, which coincide with the highest peaks.

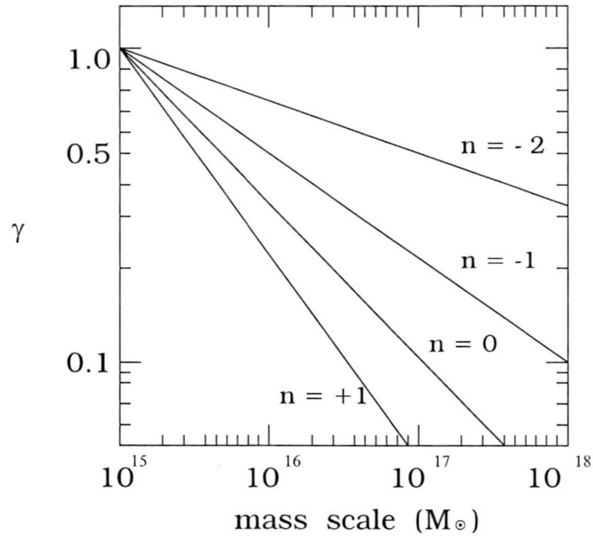

Figure 1. Correlation between cluster and large-scale perturbations. A cluster mass of $10^{15}M_o$ has been assumed, with $\gamma = 1$ by definition on that scale. From West (1994).

In order to describe the non-linear evolution of the density field these authors construct a Zel'dovich map, $\mathbf{x}(\mathbf{r}, t) = \mathbf{r} - \mathbf{s}(\mathbf{r}, t)$, from a initial state (Lagrangian) space $\mathbf{r}$ to a final state (Eulerian) space $\mathbf{x}$, where Lagrangian coordinates define the initial positions of particles and the Eulerian coordinates define their position at any arbitrary time. For the non-linear evolution, the displacement field $\mathbf{s}$ is the result of a background field flow $\mathbf{s}_a$ smoothed over a large scale R_b and a changeable field $\mathbf{s}_b$ built from shorter-wavelength components. The whole picture would correspond to following the orbits of particles which are determined both by an overall flow (in a way similar to the Hubble flow) and by their interaction with their closest neighbors. For our proposes of analyzing how the large-scale strain field enhances the primordial coherence of peaks, we shall focus on the region defined by the filtering scale R_b. Thus, one can characterize this filtering by its linear rms density fluctuation amplitude, $\sigma_\rho(R_b, t)$. In the linear approximation, the Zel'dovich map is defined by $\mathbf{s}_b(\mathbf{r}, t) = D(t)\mathbf{s}_b(\mathbf{r})$, where $D(t)$ describes the linear growth of fluctuations.

The background strain field is defined by

$$e_{b,ij}(\mathbf{r}) = -\frac{1}{2}\left(\nabla_i s_{bj} + \nabla_j s_{bi}\right) , \tag{9}$$

from which we can obtain the shear tensor as a time derivative, $\dot{e}_{b,ij}$. Since $\mathbf{s}_b$ is proportional to the smoothed gravitational potential, the smoothed tidal field on an element of matter will also be proportional to $e_{b,ij}$. The strain tensor can be diagonalized to find its main eigenvalues, $-\lambda_{\nu j}$, which can be ordered without lack of generality as $\lambda_{\nu 3} \geq \lambda_{\nu 2} \geq \lambda_{\nu 1}$. Define now the shear ellipticity and prolaticity, $\epsilon_\nu = (\lambda_{\nu 3} - \lambda_{\nu 1})/(2\nu_b\sigma_\rho), p_\nu = (\lambda_{\nu 1} + \lambda_{\nu 3} - 2\lambda_{\nu 2})/(2\nu_b\sigma_\rho)$, respectively. These two shear parameters are used for the calculation of the multiple-point correlation functions in Lagrangian space. In the first instance, the behavior of an ambient patch about an arbitrary point in the medium is considered, at which

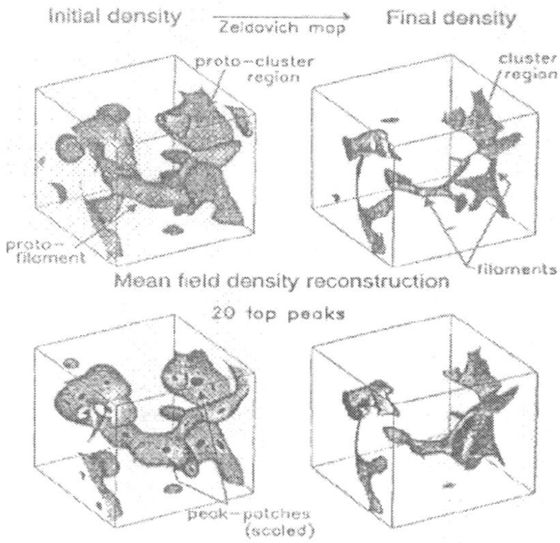

Figure 2. N-body simulations showing the formation of filamentary structures. The primordial distribution and orientation of peaks determine the filamentary structure traced by rich cluster-like structures.

point the shear parameters are expected to be typical values corresponding to a given height ν_b in the Lagrangian space. What is detected is that the initial density contours around high-σ patches trace more a filamentary than a sheet-like structure long before the Zel'dovich mapping and the density coherence scale along the filament axis are enhanced, suggesting a primordial filamentary net.

Even when these results provide support to a coherent correlation of the environment around high density peaks, it is necessary to take into account the correlations constrained by at least two or more high-σ peaks in order to show the coherence in large filamentary structures. The peak-patch theory developed by Bond & Myers (1996) has shown (Figure 2) that it is statistically likely that the shear tensor for a peak-patch will be fairly aligned with the shear tensor of neighboring high peak patches and will mimic the early stages of filamentary structures. The degree of coherence becomes weaker as the separation between peaks increases or the orientation of the shear tensor changes. The ultimate reason for the coherence in the field is the high degree of constructive interference of the density waves to form rare high peaks and to orient them.

2.3. A Coherent Primordial Density Field

Properties such as the amplitude, the shape, the orientation and the strength of the deformation tensor acting on density peaks are determined by the superposition of density waves. One alternative way to uncover the influence coming from the large-scale field on galactic-scale peaks consists of measuring changes in these properties as the density field is smoothed on increasing mass scales $M = M(R_f)$. Using a larger filtering radius, R_f, averages the generation of more massive perturbations with larger characteristic radii. In this section, *we will show that a high coherence in the primordial density field from large to small scales exists.* We have generated a primordial density field characterized by a power law spectrum, proper for numerical simulations (kindly provided by L. Moscardini and his group). Once the primordial density field is generated within a 64^3 cubic grid, and smoothed with a Gaussian filter function, the peak mass grows according to Eq. 8. A radius of $R_f^{(1)} \approx 0.6\text{Mpc} \equiv 1$ corresponds with a galactic mass scale $M \approx 10^{12} M_\odot$. For rich clusters, $R_f^{(1)} \approx 10\text{Mpc} \equiv 1$ yields $M \approx 10^{15} M_\odot$. Next, we identify the positions $\mathbf{x}_i$ of the density peaks. Suppose now, that the density field is smoothed on a larger scale by using $R_f^{(2)} = 2(\approx 1.3\text{Mpc})$. If changes in the properties of the peaks at the positions $\mathbf{x}_i$ are observed, then they can be attributed to the remaining superposition of density waves bounded by the spherical shell defined by the difference $\Delta R_f = R_f^{(2)} - R_f^{(1)}$. When $R_f^{(2)} \approx R_f^{(1)}$, the general properties of a density peaks do not change, and both scales will be correlated. For an increasing ΔR_f, slight changes in the properties of density peaks would mean a strong correlation on those scales. Hereafter, the distance between shells is specified by taking two consecutive filtering scales as $R_f^{(i+1)} - R_f^{(i)}$, where $R_f^{(8)}$ represents the largest scale used: 4.8 Mpc when we analyze the effect on galactic-scale peaks, and 80 Mpc for analyzing the effects on cluster-scale peaks.

We shall restrict ourselves to quantify the changes in; (i) the amplitude of the perturbations, and (ii) the changes in the orientation of density perturbations (calculated by the inertia tensor) relative to strength and the main axes of the deformation tensor.

Thus, we have to evaluate the angle θ subtended by the major axes associated with the unperturbed and the perturbed deformation tensor, the former of these corresponding with $R_f^{(1)}$ and the latter one, being the result of increasing the smoothing scale. The deformation tensor is calculated from the Zel'dovich (1970) approximation as

$$\mathbf{D}_{jk} = \delta_j k + b(t)\frac{\partial v_k}{\partial q_j}, \tag{10}$$

where the density and the velocity field are interpolated when necessary. The validity of the computed peculiar velocity is checked through the continuity equation (Peebles 1980). Before going into the results, we shall remark that the discussion will be addressed as a power law spectra, $P(k) \propto k^n$, with $n = 1, 0, -1, -2$. The spectrum for CDM models can be fitted through a power law with the appropriate index. For instance, the predicted shape of the CDM fluctuation spectrum corresponds to $P(k) \propto k^0 - k^1$ on subgalactic and galactic scales, and $P(k) \propto k^{-1} - k^{-2}$ on the scale of rich clusters and superclusters.

The sequence of histograms of Figure 3 show the total number N_{pk} of $R_f^{(1)}$-deformation tensors (referred as peaks in this and the following paragraph) versus $\cos\theta$ as a function of the mass scale. For all the spectral indices we observe a trend: the nearest shell produces no important changes in the orientation of the deformation tensor –as can be deduced from the small changes produced in the form of the histograms. In general, the influence on the peaks shall decrease with more distant mass shells, equivalent to larger-scale density field. One can also see that the form of the histograms changes faster for steeper spectra, than for flatter ones. To clarify this point, consider the form of the histograms for the cases $n = 1, 0$. The maximum of the distribution of peaks lies close to $\cos\theta = 1$, *i.e.*, for a significant fraction of the peaks their orientation is correlated with their local deformation tensor. The number of peaks within the interval $0 < \cos\theta < 0.9$, for which the orientation changes are larger than $\sim 25°$, is approximately $60 - 80\%$ of the total sample for $R_f^{(2)} - R_f^{(1)} \approx 1.2$ Mpc, and decreases to $10 - 20\%$ for $R_f^{(3)} - R_f^{(2)} \approx 2$ Mpc. This means that effects on the density peaks are mainly due to the neighboring distribution of mass.

For flat spectra, $n = -1, -2$, the situation is different. The form of the histograms is not drastically modified even to large filtering scales. The peak of the histograms is sharper near $\cos\theta = 1$ up to scales of ≈ 3 Mpc – *i.e.*, a large number of peaks present no changes in the orientation of the deformation tensor. Such scales nearly match the scale of a cluster of galaxies. Thus, a correlation between a cluster-scale perturbation and its central regions would be feasible. Different numerical simulations (*e.g.*, Faber, 1982; Blumenthal et al., 1984; White & Frenk, 1991) suggest that a power spectrum index of $n \approx -2$ is consistent with the observed galaxy properties. If we identify galaxies in a one-to-one mapping with peaks in the density field, our results would suggest that the orientation of the deformation tensor for galactic-scale perturbations is correlated with the surrounding mass on the scale of the parent cluster. Furthermore, the observational studies by Vogeley et al. (1992) and Einasto et al. (1993) have estimated the shape of the fluctuation spectrum on scales of $\sim 20 - 100$ Mpc, finding a spectral index $n \approx -1.5 \pm 0.5$. Which brings as a possible consequence that the degree of correlation observed in Figures 11 and 12, for flat spectra, suggests that a correlation of clusters with larger scales should be present. What this really suggests to us is that orientation of the main axes of the local tidal shear is dominated by the large-scale field.

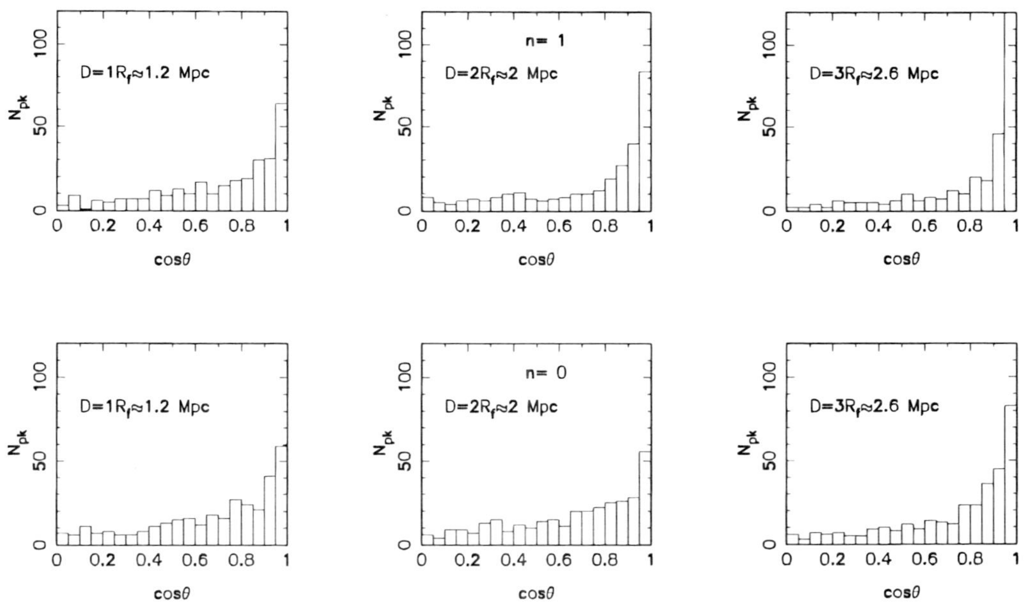

Figure 3. Histograms showing the changes in the orientation of the deformation tensor, as a function of the shell distance, for $n = 1$ and $n = 0$.

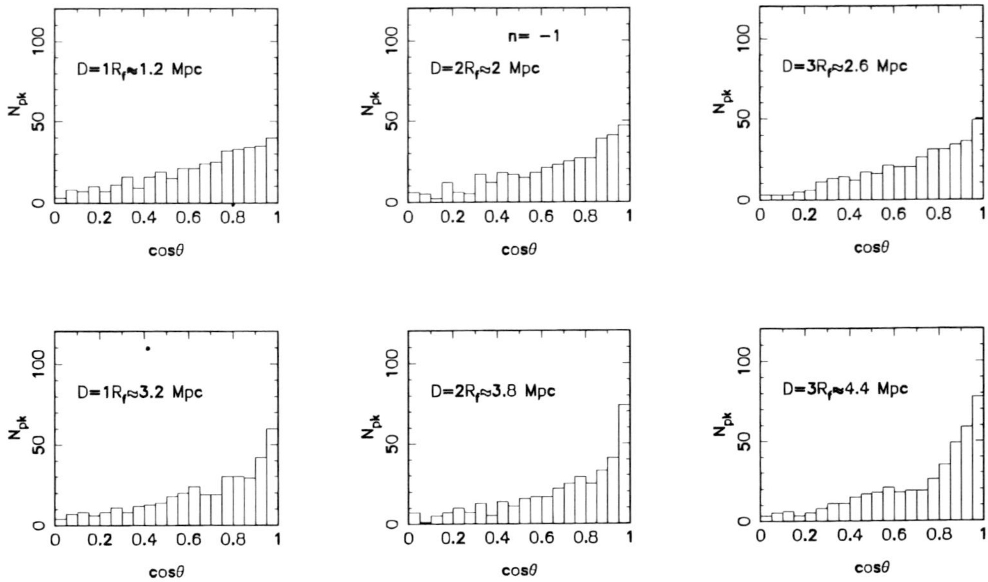

Figure 4. The same as Figure 3 for $n = -1$.

In order to analyze whether these conclusions are also satisfied by the orientation of peaks, we calculated the relative orientation between the main axes of the deformation tensor and those of the inertia tensor for $R_f^{(1)} = 0.6$ Mpc. Figure 6 displays the distribution of relative orientation as a function of the spectral index. It can be seen that there is an important fraction of the peaks whose tensors are similarly oriented; $\sim 30 - 40\%$ of the whole sample for $n = -1$ and $n = -2$, and approximately $10 - 20\%$ for $n = 1$ and $n = 0$. The misalignment of the main axes of the two tensors would imply that the final morphology of the structures formed from the collapse of perturbations will be poorly correlated with the initial shape of their progenitors. It is partially due to this misalignment that the perturbations will acquire angular momentum with no correlation between neighboring galaxies. Thus, the tidal shear effects produced by the large-scale density field will affect the evolution of peaks depending on how they are initially oriented with the tidal field.

So far, we have analyzed the changes in the orientation of the principal axes of the deformation tensor without paying attention to the magnitude of the deformation itself. We now assess such magnitudes in order to estimate the importance of the tidal effects produced on the peaks by the surrounding mass distribution. We first estimate the tidal field effects on the deformation tensor of peaks through the relation between the eigenvalues, λ_i, of the deformation tensor and the density contrast δ

$$\sum_{i=1}^{3} \lambda_i^{(j)} \propto \delta^{(j)}, \tag{11}$$

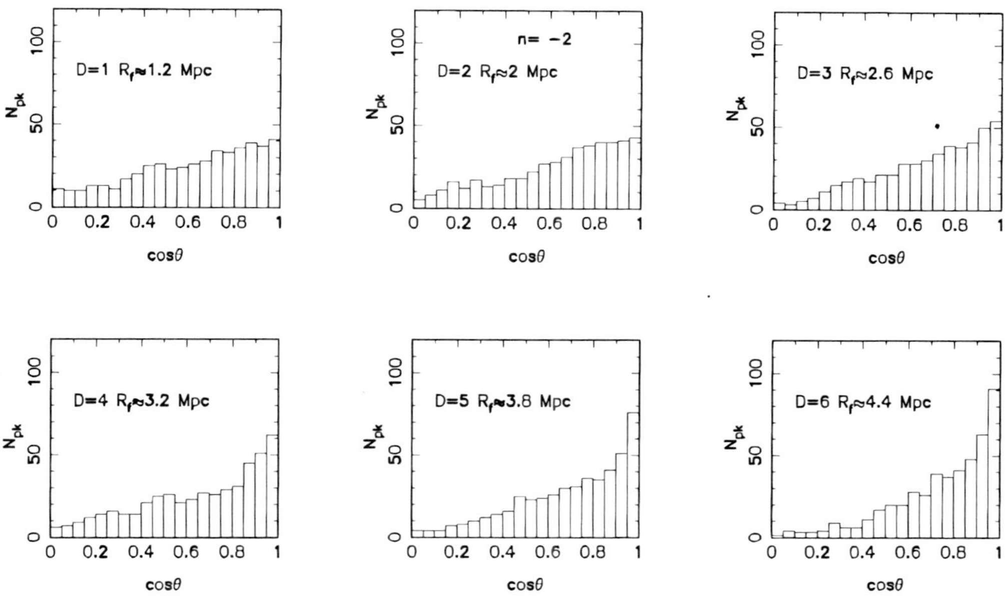

Figure 5. The same as Figure 3 for $n = -2$.

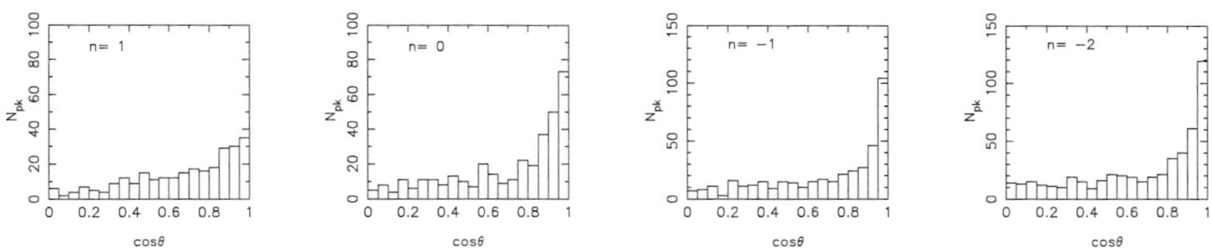

Figure 6. Histograms showing the misalignment between the main axes of the deformation tensor and those of the inertia tensor.

where j denotes the values corresponding to different mass scales. The difference

$$\sum_{i=1}^{3} \lambda_i \equiv \sum_{i=1}^{3} (\lambda_i^{(j+1)} - \lambda_i^{(j)}) \propto \delta^{(j+1)} - \delta^{(j)} , \tag{12}$$

shall provide us with a measure of the strength of the effect produced by the external field. Figure 7 shows the distribution of peaks as a function of $\sum \lambda_i$. The high number of peaks at $\sum \lambda_i \approx 0$ indicates a weak influence of the nearest mass distribution on the deformation tensor from the fact that the eigenvalues remain unperturbed. This behavior is observed for all the spectral indexes. By comparing the changes in the shape of the initial distribution of peaks, as the filtering scale increases, we do not observe important tidal effects for the cases $n = 1$ and $n = 0$. For flat spectra, however, the initial shape of the distribution is gradually modified up to scales of $\Delta R_f \approx 4 - 5 \approx 4$ Mpc. After these scales the effect decreases significantly as shown by the high number of peaks distributed around $\sum \lambda_i \approx 0$. This suggests again an influence of the large-scale superposition of density waves on the peaks. On scales for clusters of galaxies, the values $R_f \approx 4-5$ would correspond to a tidal effect of cluster-scale perturbation with their environment on scales of $\approx 30 - 40$ Mpc.

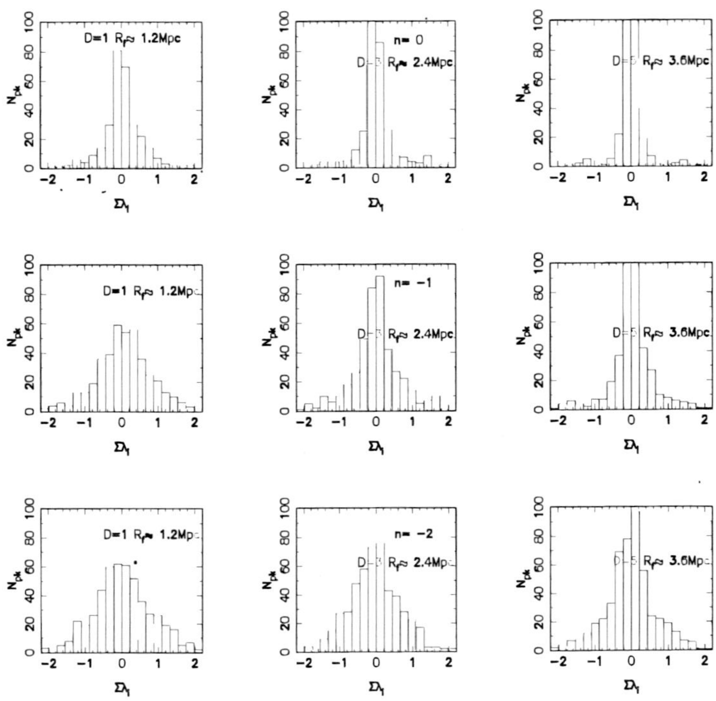

Figure 7. Histograms of $\sum_{i=1}^{3} \lambda_i = (\lambda_i^{(j+1)} - \lambda_i^{(j)})$ as a function of the shell distance. The cases $n = -1, -2$ show small changes at the distances indicated in the panels.

We now provide a further measure of the changes in the deformation tensor, by resorting to a concept widely used in the theory of matrix perturbation: the norm of the perturbation matrix (c.f. Stewart & Sun 1990). Roughly, the goal of this theory is to predict, or bound, the changes in physical processes described by a matrix when the elements of the matrix change. The theory would assess, for example, how far the '*magnitude*' of a matrix associated to a physical process – *e.g.*, the deformation tensor $\mathbf{D}_{ij}^{(1)}$– will change when the elements of the matrix change according to $\mathbf{D}_{ij}^{(2)} = \mathbf{D}_{ij}^{(1)} + \mathbf{\Delta}_{ij}$, due to a perturbation, $\mathbf{\Delta}_{ij}$. The prerequisite for answering this question is to make a precise the term for 'magnitude', which is done by defining a norm in the space of matrices. We adopted the Frobenius norm defined as

$$\mathbf{F} \equiv \| \mathbf{\Delta}_{ij} \| \equiv \sqrt{\sum_{i,j} | \alpha_{ij} |^2} \equiv \sqrt{\sum_i \lambda_i^2} . \tag{13}$$

When the deformation tensor does not suffer important changes, the norm will tend to take small values. The lower limit corresponds to $\mathbf{F} = 0$, in which the elements of the deformation tensor are not affected.

Figure 8 displays the distribution of peaks with norm **F**. A clear difference in the strength of the tidal field is observed between the different spectral indices. The maximum of the distribution of peaks, in the cases $n = -1$ and $n = -2$, shows a considerable deviation from zero at large filtering radii, thus confirming the influence of the superposition of the density waves which form the large-scale perturbations on the local deformation tensor of small-scale perturbations. This is also seen from Figure 9a, where we plot the mean strength F as a function of R_f. The norm of the deformation tensor at $R_f^{(1)}$ was used to normalize the maximum effect to unity. We confirmed our prior results: steep spectra show a tidal effect coming from the nearest distribution of mass. Flat spectra show im-

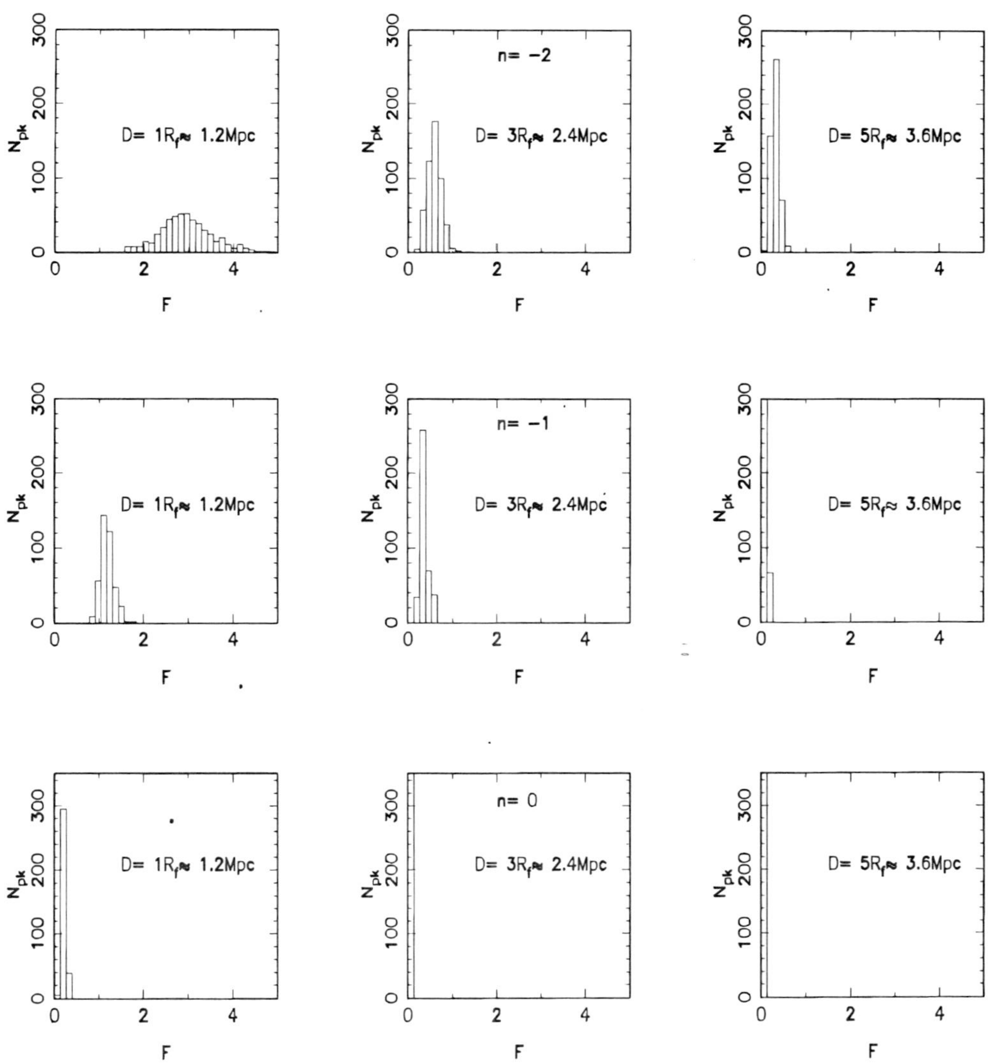

Figure 8. Histograms of the Frobenius norm, as a function of the shell distance and the spectral index. For $n = 0$, $F = 0$ suggests no important changes coming from the larger scales.

portant effects coming from even larger-scales. An insight into the decaying rate of **F** is provided by the difference of **F**'s between different shells, which we binned as

$$\Delta F_i = |F(R_f = 8/i) - F(R_f = 4/i)| \;, \quad \text{for} \quad \mathrm{i} = 4, 2, 1 \tag{14}$$

and plotted in Figure 9b. A tendency for the influence to decay is observed. However, the rate of decay is faster for those models with steep spectra. From these results, one can see that there is no divergence due to the mass increasing in the density field. However, it continues to produce important effects on small-scale peaks.

One can expect that an alignment of peaks and a galaxy-cluster-like alignment will be apparent if the orientations of the inertia tensor of peaks are compared with the orientation of their host larger fluctuations. Thus, we explored the parallel alignment possibilities by calculating the cosine of the angle between the major axis of the inertia tensor of galactic-scale peaks and the major axis of the inertia tensor of the larger-scale fluctuation.

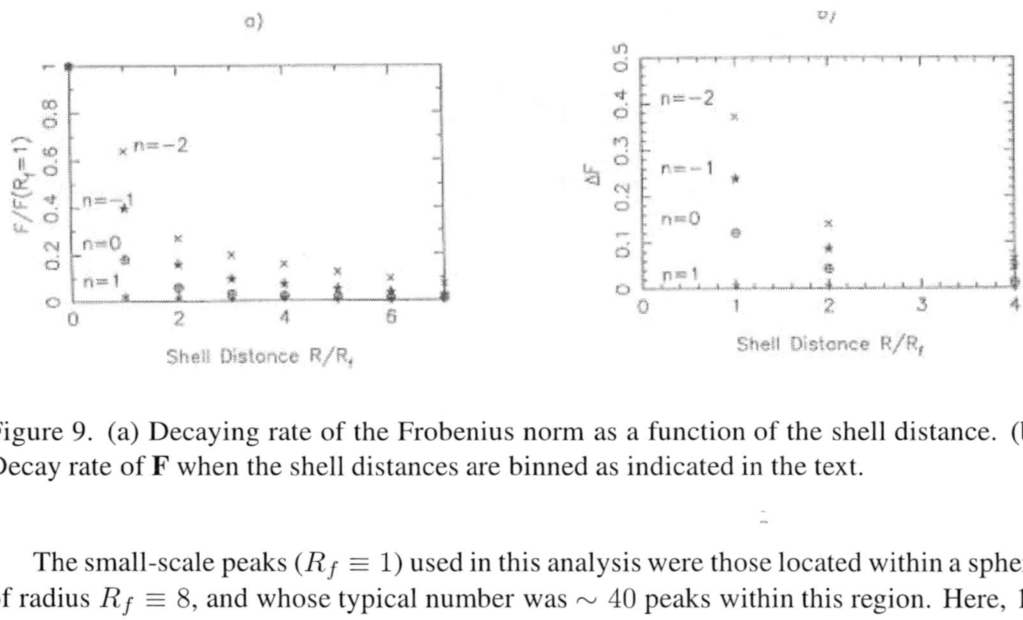

Figure 9. (a) Decaying rate of the Frobenius norm as a function of the shell distance. (b) Decay rate of **F** when the shell distances are binned as indicated in the text.

The small-scale peaks ($R_f \equiv 1$) used in this analysis were those located within a sphere of radius $R_f \equiv 8$, and whose typical number was $\sim$ 40 peaks within this region. Here, 10 cases were analyzed for each of our 6 realizations of a spectrum with $n = -2$. Some of the results are presented in Figure 10. Important alignment effects were detected in two cases – these are marked as [3] and [6]. Each of these cases includes $\approx 28\%$ as the total number of peaks. However, we could not get any clear insight into the ultimate reason for their alignment (*e.g.*, highest peaks, ellipticity, number of neighbors), as might allow us to reproduce them with a major statistical weight in a larger realization. What is noteworthy is the existence of some cases for which alignment exists.

An even weaker trend for parallel alignments showed for the other cases – which include at most 20% of the peaks – is probably not too surprising since we carried out the comparison of the orientations with peaks on a scale larger than that where we found evidence of a stronger correlation, $\Delta R_f \approx 4 - 5$. These results, therefore, are useful as indicators of an upper limit at which we could expect primordial alignment effects. If one adopts a cluster-scale smoothing of the density field, the present results would suggest that we cannot expect alignment of the cluster-cluster type on scales larger than $40 - 60h^{-1}$

Mpc, in agreement with some observational results: an alignment of the main axes of clusters of galaxies has been detected up to scales of $30 - 60h^{-1}$ Mpc (e.g., Binggeli 1982; Oort 1984; Plionis 2002, 2003) and even at the central regions of clusters between the cD galaxy and their host cluster (Rhee et al. 1992; West 1994).

For the steep spectra, the distributions showed no signals for any kind of alignment. The orientation of density peaks was uniformly distributed.

We have also found that the inertia tensor of about $10 - 15\%$ of the density peaks is oriented with the main axes of the large-scale perturbations on scales smaller than $\Delta R_f \approx 8$. Hence, one should expect that at the moment of collapse the inner regions would already be aligned with the host structure (Figure 11). The primordial tidal field acting on these perturbations will not significantly change their initial orientations because the quadrupole term of the peaks, which is the first to become coupled with the tidal field, would be already oriented in that direction.

3. Conclusion

Tidal shear has been frequently invoked to generate alignment effects. However, the successful theory of cD galaxy formation by anisotropic mergers of smaller gas-rich protogalactic units can account for the observed alignment of a cD galaxy with its host cluster major axis. These two ingredients seem to work together for moving the material in the direction dictated by the formation of large-scale filaments. Major support for this the-

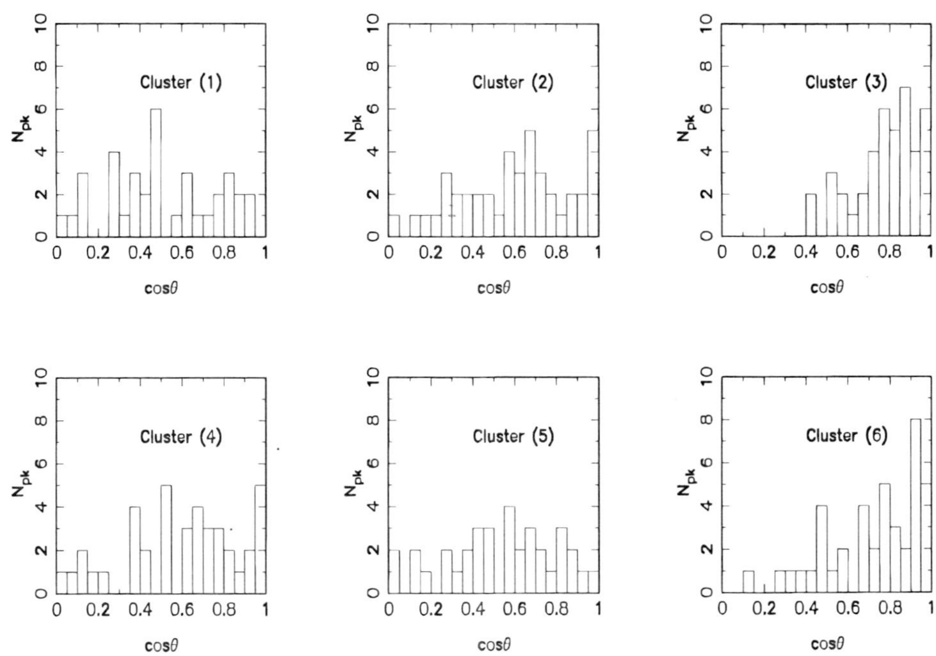

Figure 10. "Clusters of peaks" showing the orientational distribution of their members. Cases [3] and [6] show a tendency for alignments.

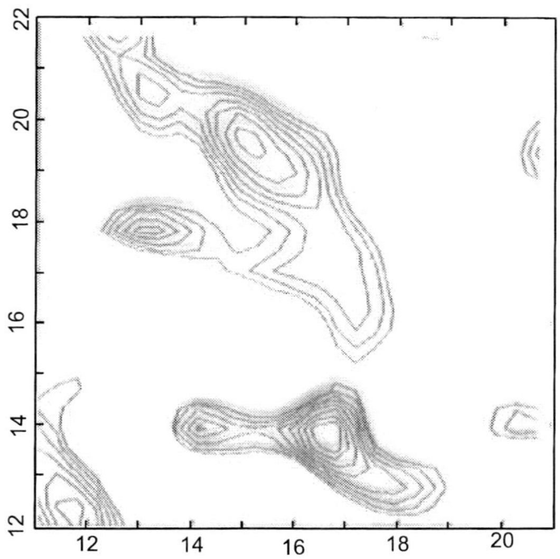

Figure 11. Example of isodensity profiles of connected density peaks with equal orientations in space.

ory is given by the primordial alignment effects detected by Bond et al. (1996) using the peak-patches theory. This primordial alignment of rare, high density peaks is, in principle, the ultimate reason for coherent large-scale filaments, in which protocluster scale perturbations are already aligned. Protofilaments evolve as the density field enters the non-linear regime, but the orientation of high density peaks remains as a fossil quantity mainly due to the anisotropic tidal field which they produce. The final product of this mechanism is the formation of galaxy bridges linking clusters. The tidal shear produced by a collapsing, large-scale anisotropic distribution of matter is the dominant process to generate the coherence from the innermost region of cD galaxies up to the alignment of clusters on scales of $50 - 70h^{-1}$ Mpc – and between clusters, higher galaxy counts along the direction defined by the major axis of cD galaxies.

The constraint imposed by requiring a high peak to be oriented along a specified axis forces the density waves that had to constructively interfere at the peak to constructively interfere also along the long axis. For flat spectra, long wavelengths as well as short wavelengths form the high peaks, hence the effect can still persist weakly out to large distances. This coherent wave bunching will produce the overdensities of clusters, galaxies and of the mass field largest along the long axis and smallest along short axis. The same effect will also tend to align the long axes of neighboring peaks so that they point toward each other. Further, the orientation of a central smaller scale peak will be correlated with the imposed cluster scale orientation, especially if the two scales are not quite different. Within the context of cosmological N-body simulations it is found that models with flat spectra on large scales are able to generate alignment effects, such as those reported between clusters of galaxies, and that the clustering and the strength of the alignment effects decreases as one consider steeper spectra. Thus, the existence of coherent orientation of clusters is a common feature in the Universe, if a flat spectrum on large-scales is reiterated.

Acknowledgments

This work was supported by CONACYT, SNI and PROMEP (México). The authors would also like to thank NovaScience for publishing this manuscript.

References

[1] Adler, R.J., 1981, *The Geometry of Random Fields*, Wiley, Chichester, New York.

[2] Atlay, G., Colberg, J.M., & Croft, R.A.C., 2006, *MNRAS*, **370**, 1422.

[3] Basilakos, S., Plionis, M., Yepes, G., Gottlober, S., & Turchanivov, V., 2006, *MNRAS*, **365**, 539.

[4] Bahcall N.A., 1987, in *Observational Cosmology*, Proceedings of the IAU Symposum, Beijing. Reidel, Dordrecht.

[5] Bond J.R., 1986, in *Galaxy Distances and Deviations from the Hubble Flow*, B.F. Madore and R.B. Tully (eds.), (Reidel, Dordrecht).

[6] Binggeli, B., Tammann, & Sandage, A., 1987, *AJ*, **94**, 251.

[7] Blumenthal, G.R., Faber, S.M., Primack, J.R., & Rees, M.J., 1984, *Nature* **311**, 517.

[8] Bond J.R., 1987a, in *Nearly Normal Galaxies*, S. Faber (ed.), (Springer Verlag, New York).

[9] Bond J.R., 1987b, in *The early Universe*, W.G. Unruh (ed.), (Reidel, Dordrecht).

[10] Bond J.R., 1987c, in *Cosmology and Particle Physics*, I. Hinchcliffe (ed.), (World Scientific, Singapore).

[11] Bond, J.R., & Myers S.T. 1996, The Peak-patch picture of cosmic catalogs. III, *ApJS*, **103**, 63.

[12] Bond, J.R., Kofman, L., & Pogosyan, D. 1996, *Nature*, **380**, 60.

[13] Einasto, J., Gramann, M., & Saar, E., & Tago, E. 1993, *MNRAS*, **260**, 705.

[14] Faber, S.M., 1982, in *Astrophysical Cosmology*, Bruck et al., (eds.), (Ponti. Acad. Sci., Vatican City), 191.

[15] González A., 1995, PhD. thesis, *Cosmological Alignment Effects*. University of Sussex, England.

[16] Hashimoto, Y., Henry, J. P., & Boehringer, H., 2008, *MNRAS*, **390**, 1562.

[17] Jones, B.J.T., van de Weygaert, R., & Aragon-Calvo, M.A., 2010, *MNRAS*, **408**, 897.

[18] Lee, J., & Evrard, A. 2007, Cluster-supercluster alignments, *ApJ*, **657**, 30.

[19] Libeskind, N.I., Frenk, C.S., Cole, S., Helly, J.C., Jenkins, A., Navarro, J.F., & Power, C., 2005, *MNRAS*, **363**, 146.

[20] Little, B., Weinberg, D.H., & Park, C., 1991, *MNRAS*, **253**, 295.

[21] Oort, J H., 1984, in *Clusters and Groups of Galaxies*, F. Mardirossian, G. Giuricin, & M. Mezzeti, (eds.),(Reidel, Trieste).

[22] Padmanabhan, T. 1999, *Structure Formation in the Universe*, Cambridge University Press, 2nd Edition.

[23] Peebles, P.J.E., 1980, *The Large-scale Structure of the Universe*, Princeton University Press.

[24] Plionis, M., & Basilakos, S., 2002, The cluster substructure-alignment connection. *MNRAS*, **329**, L47.

[25] Plionis, M., Maurogordato, S., Benoist, C., Ferrari, C., & Basilakos, S., 2003, Galaxy Alignments as a Probe of the Dynamical State of Clusters, astro-ph/0305215.

[26] Stewart, G.W., & Sun, Ji-Guang, 1990, *Matrix Perturbation Theory*, Academic Press.

[27] Rhee, G., van Haarlem, M., & Katgert, P., 1992, *AJ*, **103**, 1721.

[28] Splinter, R.J., Melott, A.L., Linn, A.M., Buck, C., & Tinker, J., 1997, The ellipticity and orientation for clusters of galaxies in N-body experiments. *ApJ*, **479**, 632.

[29] West, M.J., 1994, *MNRAS*, **268**, 79.

[30] West, M.J., Dekel, A., & Oemler, A., 1989, *ApJ*, **336**, 46.

[31] West, M.J., Jones, C., & Forman, W., 1995, *ApJ.*, **451**, L5.

[32] West, M.J., Villumsen, J., & Dekel, A., 1991, *ApJ*, **369**, 287.

[33] West, M.J., Weinberg, D.H., & Dekel, A., 1990, *ApJ*, **353**, 329.

[34] White, S.D.M., & Frenk, C.S., 1991, *ApJ*, **379**, 52.

[35] Vogeley, M.S., Park, C., Geller, M.J., & Huchra, J.P., 1992, *ApJ.* (Lett), **391**, 5.

[36] Zel'dovich, Y.B. 1970, *AAp*, **5**, 84.

In: Galaxies: Evolution, Morphology and Dynamics
Editors: A. Guerrera Sanchez et al., pp. 63-109
ISBN: 978-1-62081-185-6

Chapter 3

Exploring the Nature of Motion in Quiet and Active Galaxies with a Satellite Companion

Euaggelos E. Zotos*
Department of Physics
Section of Astrophysics, Astronomy and Mechanics
Aristotle University of Thessaloniki,
Thessaloniki, Greece

Abstract

In this chapter, we present a galactic gravitational model of three degrees of freedom, in order to study and reveal the character of the orbits of the stars, in a binary stellar system composed of a primary quiet or active galaxy and a small satellite companion galaxy. Our main dynamical analysis will be focused on the behavior of the primary galaxy. We investigate in detail the regular or chaotic nature of motion, in two different cases: (i) the time independent model in both 2D and 3D dynamical systems and (ii) the time evolving 3D model.

For the description of the structure of the 2D system, we use the classical method of the Poincaré $x - p_x, y = 0, p_y > 0$ phase plane. In order to study the structure of the phase space of the 3D system, we take sections in the plane $y = 0$ of the 3D orbits, whose initial conditions differ from the plane parent periodic orbits, only by the z component. The set of the 4-dimensional points in the (x, p_x, z, p_z) phase space are projected on the (z, p_z) plane. The maximum Lyapunov Characteristic Exponent (L.C.E) is used in order to make an estimation of the chaoticity of our galactic system, in both 2D and 3D dynamical models. In order to distinguish between ordered and chaotic motion in the 2D system, we apply the $S(g)$ dynamical spectrum, while for the 3D Hamiltonian system we use a more advanced spectral definition, that is the $S(w)$ dynamical spectrum

Our numerical calculations indicate, that the percentage of the chaotic orbits increases when the primary galaxy has a dense and massive nucleus. The presence of the dense galactic core also increases the stellar velocities near the center of the galaxy. Moreover, for small values of the distance R between the two bodies, low energy stars

*E-mail address: evzotos@astro.auth.gr

display chaotic motion, near the central region of the galaxy, while for larger values of the distance R, the motion in active galaxies is entirely regular for low energy stars. Our simulations, suggest that in galaxies with a satellite companion, the chaotic nature of motion in not only a result of the galactic interaction between the primary galaxy and its companion, but also a result caused by the presence of the dense nucleus in the core of the primary galaxy. Theoretical arguments are presented in order to support and interpret the numerically derived outcomes. Furthermore, we follow the 3D evolution of the primary galaxy, when mass is transported adiabatically from the disk to the nucleus.

Our numerical results are in satisfactory agreement with observational data obtained from the M51-type binary stellar systems. Comparison between the present research and with similar and earlier work is also made.

PACS: 00.00-a, 00.00.Mw, 00.00.Fm.

Keywords: kinematics and dynamics; binary stellar systems; interacting galaxies.

1. Introduction

It is well known that galaxies are not observed as isolated entities, but often appear in binary, triple or multiple stellar systems. On the other hand, just as there are clusters of stars, there are also clusters of galaxies. This leads to gravitational galactic interactions. The Milky Way-Magellanic Clouds system is a very indicative example of galactic interaction (see Gardiner et al., 1994; Lin, Jones & Klemola, 1995; Weinberg, 1995; 1998; Van der Marel, 2001; Putman et al., 2003; Connors et al., 2004; Beki & Chiba, 2005). Another interesting stellar system of interacting galaxies is the Andromeda galaxy M31, with its two smaller companion galaxies M32 and NGC 205. The gravitational effects of the M32 galaxy on the spiral structure of M31, was studied by Byrd (1878). This interesting triple stellar system was also investigated in an earlier work, using a self-consistent computer simulation code with revealing results (see Vozikis & Caranicolas, 1994).

One of the main reasons for the transition from ordered to chaotic motion, is the presence of massive objects in the central regions of the galaxies. Stars reaching the galactic center on highly eccentric redial orbits, can be scattered out of the galactic plane, displaying chaotic motion (see Sellwood & Moore, 1999). Furthermore, a central mass concentration can strongly perturb the stellar box orbits in elliptical galaxies, which become chaotic (Merrit, 1996). Observational indications suggesting the presence of a strong central mass concentration by a very sharply rising rotation curve. A second reason for chaotic behavior in galactic motion, is the presence of strong external perturbations (see Zotos, 2011c). Increasing the magnitude of perturbations weakens or destroys the stability of orbits, thereby, increasing the the percentage of the stochastisity in the stellar system. Resonances are also responsible for chaotic motion (see Contopoulos & Grosbøl, 1986; Cincotta, Giordano & Perez, 2006). Of course, resonances are not only responsible for the presence of chaos in galactic stellar systems, but also for the chaotic motion in the solar system (see Wisdom, 1987; Henrard & Caranicolas, 1990). The reader can find interesting information about the chaotic motion in galaxies and its connection with observational data in the work of Grosbøl (2002).

An extended and diffuse stellar halo envelops the Milky Way. Although only an extremely small fraction of the stars in the Solar neighborhood belong to this halo, they can be easily recognized by their extreme kinematics and metallicities. Stellar populations with these properties can now be followed to distances in excess of 100 kpc using luminous tracers such as RR Lyraes, blue horizontal branch stars, metal-poor giants and globular clusters (e.g. Oort, 1926; Baade, 1944; Eggen, Lynden-Bell & Sandage, 1962; Searle & Zinn, 1978; Yanny et al., 2000; Vivas & Zinn, 2006; Morrison et al., 2009).

In recent years, large samples of halo-star velocities (e.g. Morrisonet al., 2000; Starkenburg et al., 2009) and "tomographic" photometric and spectroscopic surveys have shown that the stellar halo is not a single smoothly-distributed entity, but instead a superposition of many components (Belokurov et al., 2006; Jurić et al., 2008; Bell et al., 2008; Carollo et al., 2007; 2009; Yanny et al., 2009). Notable substructures in the Milky Way halo include the broad stream of stars from the disrupting Sagittarius dwarf galaxy (Ibata, Gilmore & Irwin, 1994; Ibata et al., 2001), extensive and diffuse overdensities (Jurić et al., 2008; Belokurov et al., 2007a; Watkins et al., 2009), the Monoceros "ring" (Newberg et al., 2002; Ibata et al., 2003; Yanny et al., 2003), the orphan stream (Belokurov et al., 2007b) and other kinematically cold debris (Schlaufman et al., 2009). Many of these features remain unclear. At least two kinematically distinct "smooth" halo components have been identified from the motions of stars in the Solar neighborhood, in addition to one or more "thick disc" components (Carollo et al., 2009). Although current observations only hint at the gross properties of the halo and its substructures, some general properties are well-established: the halo is extensive (> 100 kpc), metal-poor ($\langle [Fe/H] \rangle \sim -1.6$, e.g. Laird et al., 1988; Carollo et al., 2009) and contains of the order of $0.1 - 1\%$ of the total stellar mass of the Milky Way (recent reviews include Freeman & Bland-Hawthorn, 2002; Helmi, 2008).

Low surface-brightness features seen in projection around other galaxies aid in the interpretation of the Milky Ways stellar halo, and vice versa. Diffuse concentric "shells" of stars on 100 kpc scales around otherwise regular elliptical galaxies have been attributed to accretion events (e.g. Schweizer, 1980; Quinn, 1984). Recent surveys of M31 (e.g. Ferguson et al., 2002; Kalirai et al., 2006; Ibata et al., 2007; McConnachie et al., 2009) have revealed an extensive halo (to ~ 150 kpc) also displaying abundant substructure. The surroundings of other nearby Milky Way analogues are now being targeted by observations using resolved star counts to reach very low effective surface brightness limits, although as yet no systematic survey has been carried out to sufficient depth (e.g. Zibetti & Ferguson, 2004; McConnachie et al., 2006; de Jong, Radburn-Smith & Sick, 2008; Barker et al., 2009; Ibata, Mouhcine & Rejkuba, 2009). A handful of deep observations beyond the Local Group suggest that stellar haloes are ubiquitous and diverse (e.g. Sackett et al., 1994; Shang et al., 1998; Malin & Hadley 1999; Martínez-Delgado et al., 2008; 2009; Faúndez-Abans et al., 2009).

Stellar haloes formed from the debris of disrupted satellites are a natural byproduct of hierarchical galaxy formation in the ΛCDM cosmology. The entire assembly history of a galaxy may be encoded in the kinematics, metallicities, ages and spatial distributions of its halo stars. Even though these stars constitute a very small fraction of the total stellar mass, the prospects are good for recovering such information from the haloes of the Milky Way, M31 and even galaxies beyond the Local Group (e.g. Johnston, Hernquist & Bolte, 1996; Helmi & White, 1999). In this context, theoretical models can provide useful "blueprints" for interpreting the great diversity of stellar haloes and their various sub-components, and for relating these components to fundamental properties of galaxy formation models. Alongside idealized models of tidal disruption, *ab initio* stellar halo simulations in realistic cosmological settings are essential for direct comparison with observational data.

In principle, hydrodynamical simulations are well-suited to this task, as they incorporate the dynamics of a baryonic component self-consistently. However, many uncertainties remain in how physical processes such as star formation and supernova feedback, which act below the scale of individual particles or cells, are implemented in these simulations. The computational cost of a single state-of-the-art hydrodynamical simulation is extremely high. This cost severely limits the number of simulations that can be performed, restricting the freedom to explore different parameter choices or alternative assumptions within a particular model. The computational demands of hydrodynamical simulations are compounded in the case of stellar halo models, in which the stars of interest constitute only $\sim 1\%$ of the total stellar mass of a Milky Way-like galaxy. Even resolving the accreted dwarf satellites in which a significant proportion of these halo stars may originate is close to the limit of current simulations of disc galaxy formation. To date, few hydrodynamical simulations have focused explicitly on the accreted stellar halo (recent examples include Bekki & Chiba, 2001; Brook et al., 2004; Abadi et al., 2006 and Zolotov et al., 2009). In the wider context of simulating the "universal" population of galaxies in representative (≥ 100 Mpc^3 cosmological volumes, these practical limitations of hydrodynamical simulations have motivated the development of a powerful and highly successful alternative, which combines two distinct modeling techniques: well-understood high-resolution N-body simulations of large-scale structure evolved self-consistently from ΛCDM initial conditions and fast, adaptable semi-analytic models of galaxy formation with very low computational cost per run (Kauffmann, Nusser & Steinmetz, 1997; Kauffmann et al., 1999; Springel et al., 2001; Hatton et al., 2003; Kang et al., 2005; Springel et al., 2005; Bower et al., 2006; Croton et al., 2006; De Lucia et al., 2006).

Therefore, it seems very challenging to construct a galactic gravitational model of three degrees of freedom (3D), in order to study and reveal the character of the orbits of the stars, in a binary stellar system composed of a primary quiet or active galaxy and a small satellite companion galaxy. The aim of the present paper is to study the character (regular or chaotic) of motion in the primary galaxy (hereafter the galaxy), under the perturbation caused by a satellite galaxy and to connect the degree of chaos with parameters such as the mass of the nucleus of the galaxy or the distance between the two galaxies. We shall also study the behavior of the velocities near the central region and try to connect this behavior with the scale length of the nucleus. Furthermore, we shall deal with the case, where adiabatic mass transfer takes place in the galaxy, that is, the galaxy evolves with time. In this initial research, only the case where the two bodies move in circular orbits about the center of mass of the system is considered, for simplicity. We plan to study the case of the satellite in inclined, elliptic orbits in a future work.

The present chapter is organized as follows: In Section 2 we present our gravitational dynamical model, which describes the motion in a binary stellar system. In Section 3 we provide an analysis of the 2D system, considering orbits in the galactic plane ($z = 0$). In the next Section, we study the character of motion in the 3D system, using different kinds of dynamical methods. In Section 5 we present an interesting theoretical analysis in order to support the numerically obtained results. In the following Section, we use a 3D time dependent model, in order to follow the evolution of orbits as mass is transported adiabatically from the disk to the nucleus of the primary galaxy. We close this chapter with Section 7, where the conclusion is presented and a comparison between the present theoretical results with observational data is made. In the same section we include statements concerning the astrophysical relevance of this research work.

2. Presentation of the Dynamical Model

Our gravitational model describes the dynamical properties of a binary stellar system, which composed of a primary galaxy and a small satellite companion galaxy. The primary galaxy (hereafter the galaxy), consists of a disk and a spherical nucleus and it is described by the potential

$$V_P(r,z) = V_d(r,z) + V_n(r,z) = -\frac{M_d}{\sqrt{b^2 + r^2 + \left(a + \sqrt{h^2 + z^2}\right)^2}} - \frac{M_n}{\sqrt{r^2 + z^2 + c_n^2}}, \tag{1}$$

where $r^2 = x^2 + y^2$, M_d, M_n is the mass of the disk and the nucleus of primary galaxy respectively, a is the disk's scale length, h is the disk's scale height, b is the core radius of the disk-halo, while c_n is the scale length of the nucleus. The disk of the primary galaxy is represented by the well known Miyamoto–Nagai model (Miyamoto & Nagai, 1975). The Plummer sphere we choose to describe the nucleus, has been used many times in the past, in order to study the effects of the introduction of a central mass component in the core of a galaxy (see Hasan & Norman, 1990; Hasan et al., 1993). The satellite galaxy is described by a point mass potential

$$V_S(r,z) = -\frac{M_s}{d}, \tag{2}$$

where $d^2 = x^2 + y^2 + z^2$, while M_s is the mass of the satellite galaxy.

In our study we shall use the theory of the circular restricted three body problem (see Caranicolas & Papadopoulos, 2009; Caranicolas & Zotos, 2009). The two bodies move in circular orbits in an inertial frame OXYZ, with the origin at the center of mass of the system, with a constant angular frequency $\Omega_p > 0$, given by Kepler's third law

$$\Omega_p = \sqrt{\frac{GM_t}{R^3}}, \tag{3}$$

where $M_t = M_d + M_n + M_s$ is the total mass of the binary system, while R is the distance between the centers of the two galaxies. A clockwise, rotating frame Oxyz, is used with axis Oz coinciding with the axis OZ and the axis Ox coinciding with the straight line joining the two bodies. In this frame, which rotates with angular frequency Ω_p, the two galactic centers have fixed positions $C_1(x,y,z) = (x_1,0,0)$ and $C_2(x,y,z) = (x_2,0,0)$ respectively. The total potential which is responsible for the motion of a star in the dynamical system of the binary stellar system is

$$\Phi_t(x,y,z) = \Phi_P(x,y,z) + \Phi_S(x,y,z) + \Phi_{rot}(x,y), \tag{4}$$

where

$$\begin{aligned}
\Phi_P(x,y,z) &= -\frac{M_d}{\sqrt{b^2 + r_{a1}^2 + \left(a + \sqrt{h^2+z^2}\right)^2}} - \frac{M_n}{\sqrt{r_1^2 + c_n^2}}, \\
\Phi_S(x,y,z) &= -\frac{M_s}{r_2}, \\
\Phi_{rot}(x,y) &= -\frac{\Omega_p^2}{2}\left[\frac{M_s}{M_t} r_{a2}^2 + R_s r_{a1}^2 - R^2 \frac{M_s}{M_t} R_s\right], \qquad (5)
\end{aligned}$$

and

$$r_{a1}^2 = (x-x_1)^2 + y^2, \quad r_{a2}^2 = (x-x_2)^2 + y^2, \quad r_1^2 = r_{a1}^2 + z^2, \quad r_2^2 = r_{a2}^2 + z^2, \qquad (6)$$

with

$$x_1 = -\frac{M_s}{M_t} R, \quad x_2 = R\left(1 - \frac{M_s}{M_t}\right) = R + x_1, \quad R_s = 1 - \frac{M_s}{M_t}. \qquad (7)$$

The angular frequency Ω_p is calculated as follows: The two bodies move around their common mass center of the system with angular frequencies Ω_{p1} and Ω_{p2}, given by

$$\begin{aligned}
\Omega_{p1} &= \sqrt{\frac{1}{x_1}\left(\frac{-dV_S(r)}{dr}\right)_{r=R}}, \\
\Omega_{p2} &= \sqrt{\frac{1}{x_2}\left(\frac{dV_P(r)}{dr}\right)_{r=R}}. \qquad (8)
\end{aligned}$$

As the primary galaxy can not be considered as a mass point, the two angular frequencies are not equal, in general. However, this issue can easily be resolved. The angular frequencies of the two bodies, can become equal under reasonable assumptions. This is justified, if the final set of the parameters has physical meaning and represents satisfactorily the dynamical system. The equation $\Omega_{p1} = \Omega_{p2}$ leads to an fourfold infinity of solutions in the four unknowns (a, b, h, c_n). If one chooses proper values (representing the dynamical system), lets say for the three parameters (b, h, c_n), then equation $\Omega_{p1} = \Omega_{p2}$ gives only two values for the parameter a. One value is positive, while the other is negative and is rejected. The author would like to make clear that, after choosing properly the parameters, the deviation between the two angular frequencies is negligible, so that $\nu = \left(|\Omega_{p1} - \Omega_{p2}|\right)/\Omega_{p1}$ is of order of 10^{-8} or even smaller and $\xi = |\Omega_{p1} - \Omega_p|$ or $\xi = |\Omega_{p2} - \Omega_p|$ is of order of 10^{-6}. Therefore, we consider the two angular frequencies practically equal, that is $\Omega_{p1} = \Omega_{p2} = \Omega_p$. Moreover, the treatment of large bodies with spherical symmetry as mass points is very common in celestial modeling. This method is quite familiar to those measuring the masses of disk galaxies. The fact that the primary galaxy, is sufficiently apart from its satellite, allow us to assume that the tidal phenomena are very small and therefore negligible.

In this rotating frame the equations of motion are

$$\begin{aligned}
\ddot{x} &= -\frac{\partial \Phi_t}{\partial x} - 2\Omega_p \dot{y}, \\
\ddot{y} &= -\frac{\partial \Phi_t}{\partial y} + 2\Omega_p \dot{x}, \\
\ddot{z} &= -\frac{\partial \Phi_t}{\partial z}, \qquad (9)
\end{aligned}$$

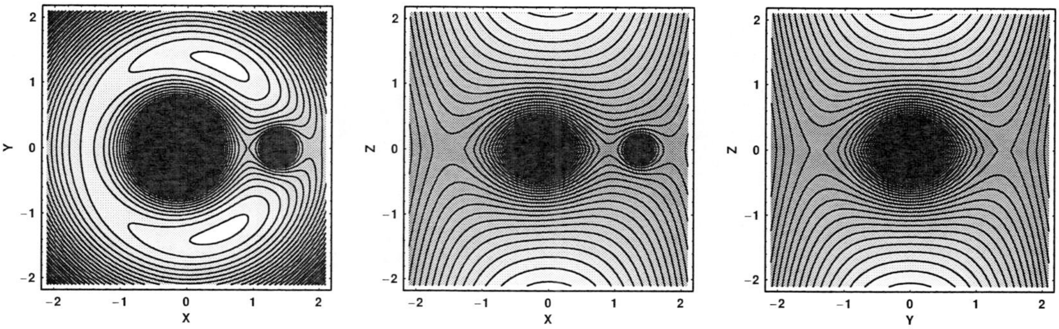

Figure 1. (a-c): Contours of the projections of the iso-potential curves $\Phi_t(x, y, z) = E_J$ on the (x, y), (x, z) and (y, z) planes.

where the dot indicates derivative with respect to the time. The only integral of motion for the system of differential equations (9), is the Jacobi integral given by the equation

$$J = \frac{1}{2}\left(p_x^2 + p_y^2 + p_z^2\right) + \Phi_t(x, y, z) = E_J\,, \tag{10}$$

where p_x, p_y and p_z are the momenta per unit mass conjugate to x, y and z, while E_J is the numerical value of the Jacobi integral.

All numerical outcomes of the present work, are based on the numerical integration of the equations of motion (9), which was made using a Bulirsh-Stöer routine in Fortran 95, with double precision in all subroutines. The accuracy of the calculations was checked by the consistency of the Jacobi integral (10), which was conserved up to the eighteenth significant figure.

In this article, we shall use a system of galactic units, where the unit of length is 20 kpc, the unit of mass is 1.8×10^{11} $M_\odot$ and the unit of time is 0.99×10^8 yr. The velocity unit is 197 km/s, while G is equal to unity (see Vozikis & Caranicolas, 1992). In these units, we use the values: $a = 0.15$, $b = 0.368$, $h = 0.00625$, $c_n = 0.0125$ and $M_s = 0.2$. The values of the above quantities of the dynamical stellar system remain constant during this research, while the values of M_d, M_n and R are treated as parameters. The above numerical values of the constant dynamical quantities of the system, secure positive density everywhere and free of singularities.

Figure 1a-c shows the contours of the projections of the iso-potential curves $\Phi_t(x, y, z) = E_J$, on the (x, y), (x, z) and (y, z) planes respectively. The values of the parameters are: $M_d = 2$, $M_n = 0.08$, $R = 1.5$ and $\Omega_p = 0.821922$. Lighter colors indicate higher values of E_J. Figure 2 shows the contours of the iso-potential curves $\Phi_t(x, y, z) = E_J$, on the (x, y) plane, with some additional details regarding the stability and the structure of the dynamical stellar system. Moreover, L_1, L_2, L_3, L_4 and L_5 are the five Lagrange equilibrium points, while C_1 and C_2 are the centers of the two galaxies at a distance $R = 1.5$. At these equilibrium points we have

$$\frac{\partial \Phi_t(x, y, z)}{\partial x} = \frac{\partial \Phi_t(x, y, z)}{\partial y} = \frac{\partial \Phi_t(x, y, z)}{\partial z} = 0\,. \tag{11}$$

L_1, L_2, L_3 are the unstable saddle equilibrium points, while L_4 and L_5 are the triangular points (see Binney & Tremaine, 2008).

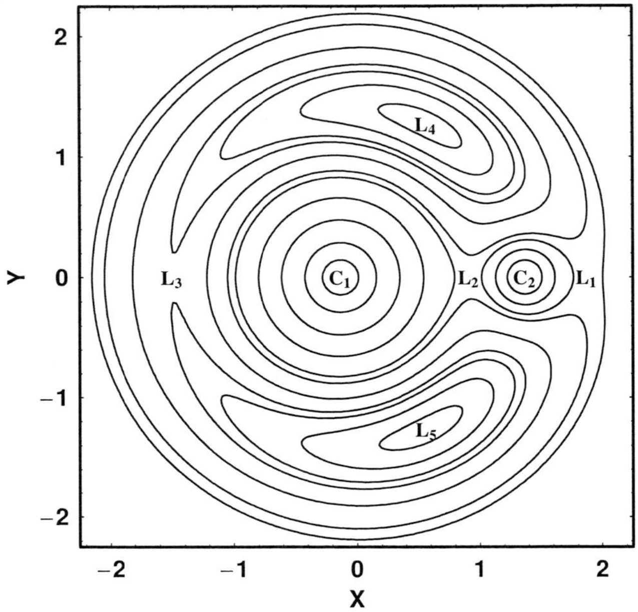

Figure 2. Contours of the projections of the iso-potential curves $\Phi_t(x, y, z) = E_J$ on the (x, y) plane. The five Lagrange equilibrium points are indicated as: L_1, L_2, L_3, L_4 and L_5, while C_1 and C_2 are the centers of the two galaxies at a distance $R = 1.5$.

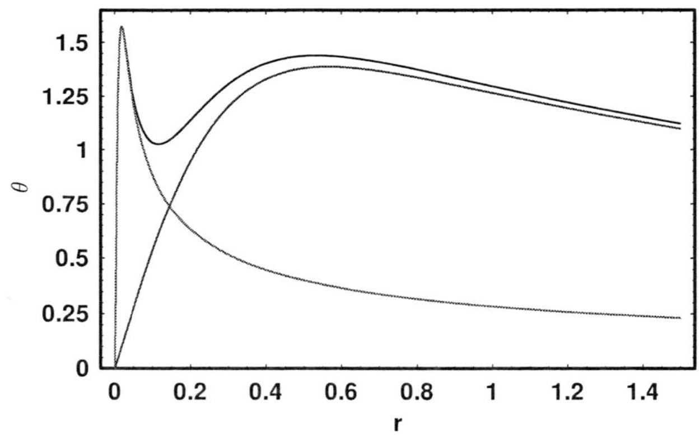

Figure 3. The rotation curve of the primary galaxy is shown as the black line. The red line is the contribution from the spherical nucleus, while the blue line is the contribution from the disk-halo potential. (Colors available only in electronic form).

The rotation curve of the primary galaxy when $M_d = 2$ and $M_n = 0.08$, is shown as the black line in Figure 3. In the same plot, the red line is the contribution from the spherical nucleus, while the blue line is the contribution from the disk-halo component. We observe, that at small distances from the primary galactic center $r \leq 0.1$ length units, which equals to 2 kpc, dominates the contribution from the spherical nucleus, while at larger distances $r > 2$ kpc the disk-halo contribution is the dominant factor.

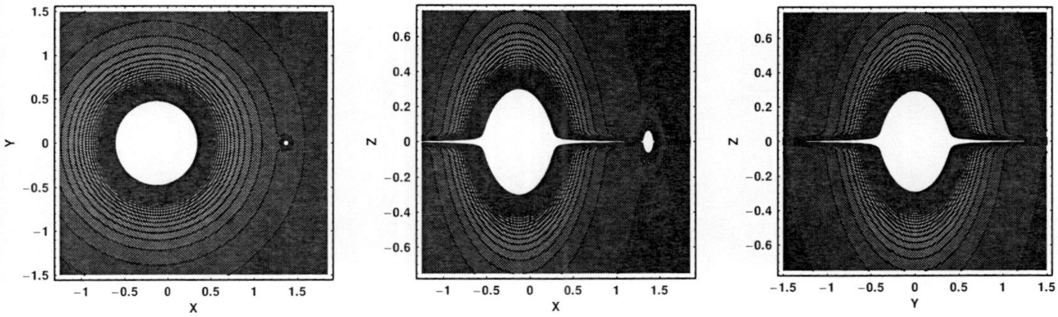

Figure 4. (a-c): Contours of the projections of the iso-density curves $\rho(x, y, z) = const$ on the (x, y), (x, z) and (y, z) planes.

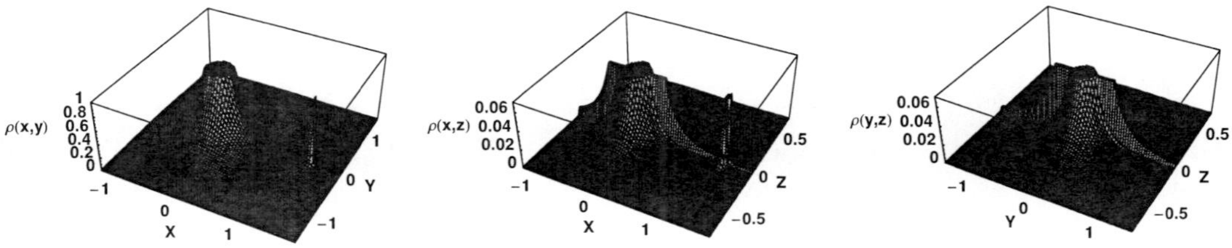

Figure 5. (a-c): The 3D plots of the distribution of the total mass density $\rho(x, y, z)$ on the (x, y), (x, z) and (y, z) planes.

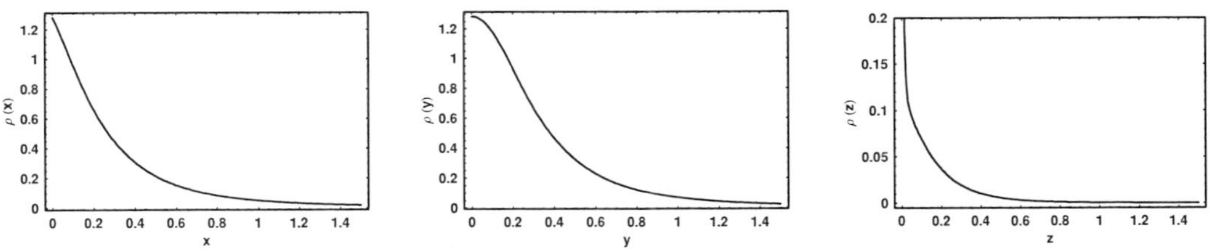

Figure 6. (a-c): The evolution of the total mass density along the x, y and z axis.

It would be very useful to compute the mass density $\rho(x, y, z)$ of the primary galaxy, derived from the potential (1), using the Poisson's equation

$$\rho(x, y, z) = \frac{1}{4\pi G}\nabla^2 V_P(x, y, z) = \frac{1}{4\pi G}\left(\frac{\partial^2 V_p}{\partial x^2} + \frac{\partial^2 V_p}{\partial y^2} + \frac{\partial^2 V_p}{\partial z^2}\right). \quad (12)$$

Figure 4a-c shows the projections of the iso-density curves $\rho(x, y, z) = const$, on the (x, y), (x, z) and (y, z) planes respectively, when: $M_d = 2$, $M_n = 0.08$, $R = 1.5$ and $\Omega_p = 0.821922$. In the first two plots, we can observe the small variations on density distribution, caused by the mass of the satellite galaxy. Figure 5a-c, depicts the 3D plots of the values of the total mass density $\rho(x, y, z)$ on the same basic planes. We can distinguish the spherical nucleus at the central region of the primary galaxy, the well formed disk structure and also the abrupt peak of the mass density at the position of the satellite galaxy ($x \simeq 1.36$). These 3D plots

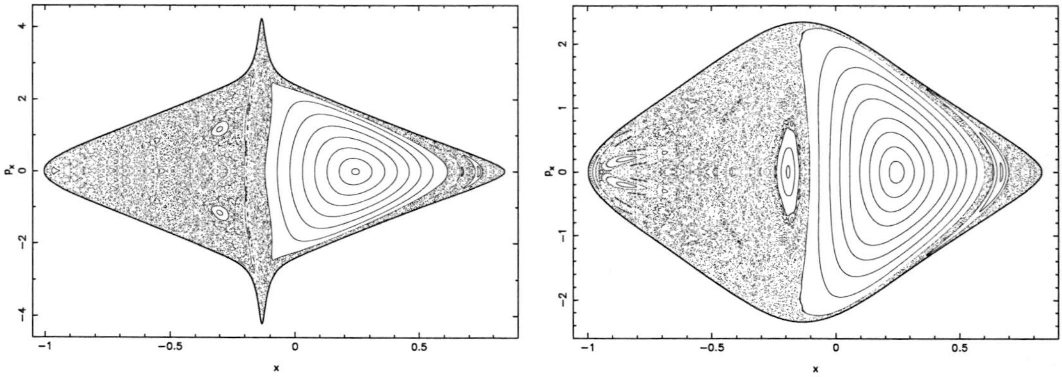

Figure 7. (a-b): The $x - p_x$ Poincaré phase planes when $R = 1.5$ and $\Omega_p = 0.821922$. (a-left): $M_d = 2$ and $M_n = 0.08$, while (b-right): $M_d = 2.08$ and $M_n = 0$.

assure that the mass density of our binary stellar system, has positive value everywhere. Going one step further, we present in Figure 6a-c the evolution of the mass density along the x, y and z axes respectively. One can observes, that in all cases the value of the mass density reduces and tends asymptotically to zero, as the distance from the center of the primary galaxy increases. Furthermore, we see that along the z axis the reduction of the mass density is more rapidly than in the two other axes. The values of the mass density ρ, on the vertical axis are in $M_\odot/pc^3$.

3. Structure of the 2D Hamiltonian System

In this section, we shall investigate the properties of motion in the Hamiltonian system of two degrees of freedom (2D). This can be derived from equation (10), if we set $z = p_z = 0$. Then the corresponding Hamiltonian writes

$$J_2 = \frac{1}{2}\left(p_x^2 + p_y^2\right) + \Phi_t(x, y) = E_{J2}\,, \tag{13}$$

where E_{J2} is the numerical value of J_2. As the dynamical system is now 2-dimensional, we can use the classical, qualitative method of plotting the successive intersections of the 2D orbits, using the $x - p_x$, $y = 0$, $p_y > 0$ Poincaré surface of section (P.S.S), in order to determine the regular or chaotic character of motion. This method has been extensively applied to Hamiltonian systems with two degrees of freedom (2D), as in these systems the Poincaré surface of section is a two dimensional plane. The results obtained from the study of the 2D system will be used, in order to help us to understand the structure of the more complicated phase space of the 3D system, which will be presented in the following section.

Figure 7a-b shows the $x - p_x$, $y = 0$, $p_y > 0$ Poincaré phase plane, for the motion of a star in the primary galaxy, which obtained by numerical integration of motion of the equations of motion (9). In this case the distance between the centers of the two galaxies is $R = 1.5$ and $\Omega_p = 0.821922$. The value of the Jacobi integral is $E_{J2} = -2.6$. In Figure 7a we have $M_d = 2$ and $M_n = 0.08$, that is, the primary galaxy has a dense and massive nucleus. We observe a large chaotic area, while the regular region is mainly confined around the stable

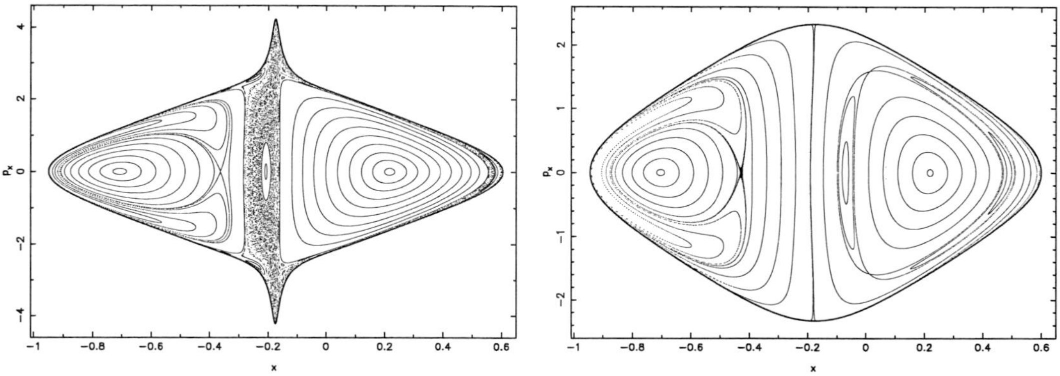

Figure 8. (a-b): Similar to Figure 7a-b but when $R = 2$ and $\Omega_p = 0.533854$.

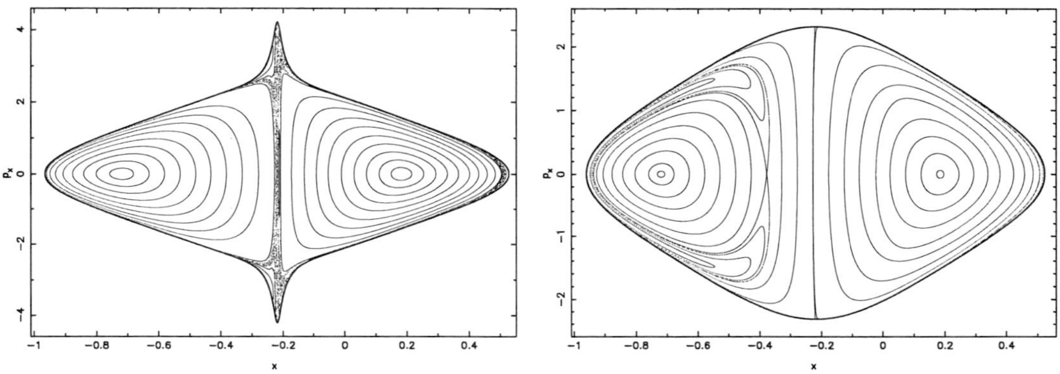

Figure 9. (a-b): Similar to Figure 7a-b but when $R = 2.5$ and $\Omega_p = 0.381995$.

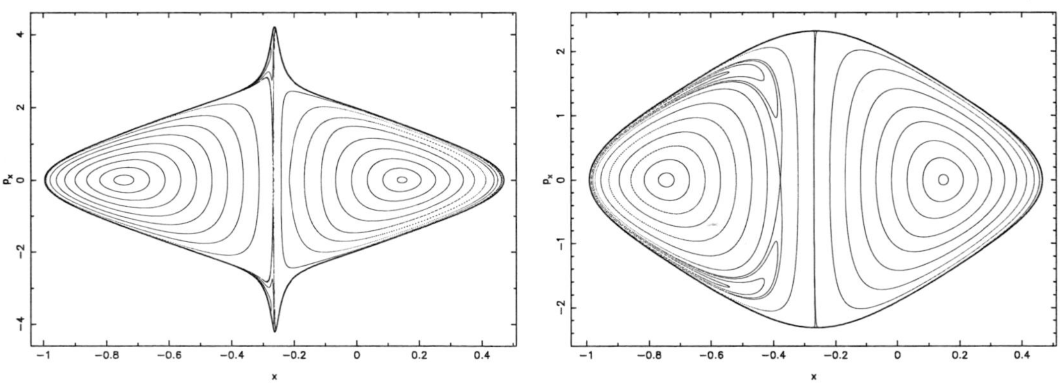

Figure 10. (a-b): Similar to Figure 7a-b but when $R = 3$ and $\Omega_p = 0.290593$.

retrograde periodic point. There are also some small islands of invariant curves embedded in the chaotic sea, corresponding to secondary resonant orbits. Note, that near the center of the primary galaxy, the velocity obtains high values of order of 900 km/s. This results from the presence of the dense nucleus in the core of the galaxy and it is characteristic of strong galactic activity. Figure 7b is similar to Figure 7a, but when $M_d = 2.08$ and $M_n = 0$. As the total mass of the system is conserved, this means that now the disk occupies all the available mass of the primary galaxy. In this case, the pattern has two main differences from the pattern shown in Figure 7a. The first difference is that the chaotic region is smaller and the second is that the velocity near the galactic core is smaller, about 470 km/s, compared to that of Figure 7a. Moreover, some secondary resonances appear in the area of the retrograde orbits. Some small sticky regions are also present in both cases (active or quiet galaxy). Therefore, one can say, that the structure of chaos in galaxies, is not only a result of galactic interaction, but it is also a result of nuclear galactic activity. In other words, the presence of the dense and massive nucleus in the galactic core of the primary galaxy, affects drastically the character of motion of the stars.

Figure 8a is similar to Figure 7a but when the distance between the centers of two the galaxies is $R = 2$ and $\Omega_p = 0.533854$. The majority of orbits around the direct and retrograde periodic point is ordered. There is a significant chaotic area, mainly near the central region of the galaxy. High velocity value is once more observed, near the galactic core of the primary galaxy, while some secondary resonances are also present. Figure 8b is similar to Figure 8a but when the galaxy is quiet, that is, when $M_n = 0$. Here, the secondary resonances look more prominent, while only one considerable small and confined chaotic layer around the direct periodic point is present. Furthermore, as the massive and dense nucleus is absent, the velocity near the center of the galactic core is smaller than that observed in Figure 8a. Figure 9a-b is similar to Figure 8a-b but when the distance between the centers of the two galaxies is $R = 2.5$ and $\Omega_p = 0.381995$. In Figure 9a, we observe that there is a relatively small chaotic layer, confined mainly near the central region of the galaxy, while the rest of the phase plane is covered by invariant curves correspond to regular orbits circulating around the stable direct and retrograde periodic point. Figure 9b is similar to Figure 9a, but when the galaxy is quiet. All orbits seem to be regular. Chaotic motion was not observed and, if present, is negligible. Figure 10a-b is similar to Figure 9a-b but when the distance between the centers of the two galaxies is $R = 3$ and $\Omega_p = 0.290593$. In Figure 10a, we observe that entire phase plane is covered by invariant curves correspond to regular orbits circulating around the stable direct and retrograde periodic point, while there is no indication of chaotic motion. Figure 10b is similar to Figure 10a, but when the galaxy is quiet. Once more, all orbits are regular and chaotic motion was not observed. Therefore, we may conclude that the primary galaxy, for large separations from its companion, do not show chaotic motion whether the dense and massive nucleus is present or not in the galactic core. The outermost solid curve in the $x - p_x$ phase planes shown in Figures 7–10 is the limiting curve, in each case.

Figure 11 shows the percentage $A\%$ of the $x - p_x$ phase plane covered by chaotic orbits as a function of the distance between the centers of the two galaxies, in two different cases. In the first case, the primary galaxy ia active and the values of the parameters are $M_d = 2$ and $M_n = 0.08$, while in the second case the primary galaxy is quiet and therefore, the values of the parameters are $M_d = 2.08$ and $M_n = 0$. The value of the Jacobi integral is always $E_{J2} = -2.6$ and the value of the angular frequency Ω_p is calculated for each particular value of the distance R from equation (3). The range of values regarding the distance between the centers of the two galaxies is $1.5 \leq R \leq 3$. We observe from Figure 11, that the chaotic percentage $A\%$ decreases exponentially, as the distance R increases in both cases (active and quiet primary galaxy). Moreover, both fitting curves converge to zero, when $R = 3$. The lower fitting curve, which corresponds to the case when the primary

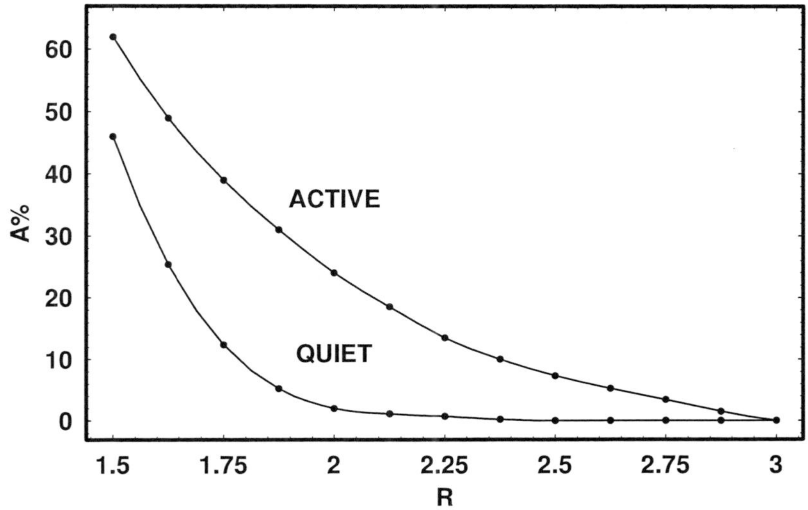

Figure 11. A plot of the area $A\%$ of the $x - p_x$ phase plane covered by chaotic orbits as a function of the distance R, when the primary galaxy is active or quiet.

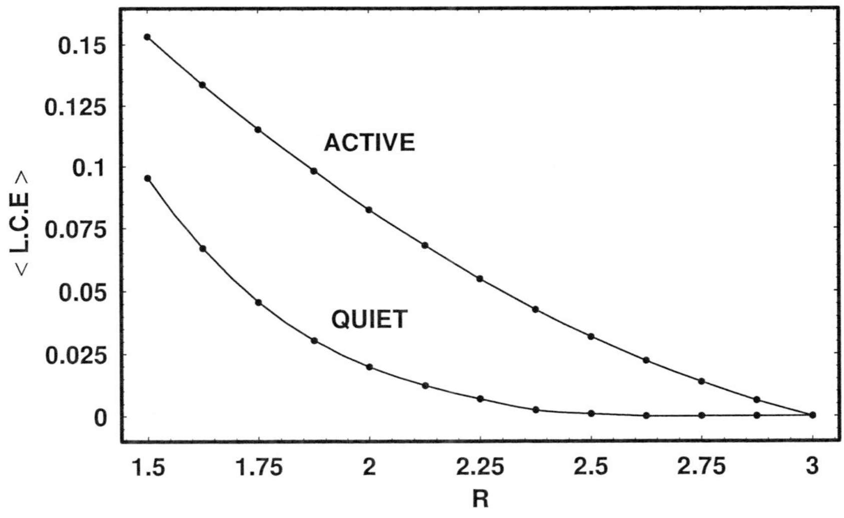

Figure 12. A plot of the average value of the Lyapunov Characteristic Exponent ⟨L.C.E⟩ as a function of the distance R, when the primary galaxy is active or quiet, for the 2D dynamical system.

galaxy is quiet, presents a more rapid reduction and reaches more quickly (when $R \geq 2.35$) to the zero limit. A more detailed view of Figure 11, reveals that for each particular value of the distance R, the chaotic percentage $A\%$ is always smaller when the dense and massive nucleus is not present in the core of the primary galaxy. We must point out, that the chaotic percentage $A\%$ is calculated as follows: we choose 1000 orbits with different and random initial conditions (x_0, p_{x0}) in each phase plane and then divide the number of those who correspond to chaotic orbits to the total number of the tested orbits.

In order to have a better estimation about the degree of chaos in our dynamical system in every case, we have computed the the average value of the Lyapunov Characteristic Exponent - L.C.E (see Lichtenberg & Lieberman, 1992). Figure 12 shows a plot of the ⟨L.C.E⟩ as a function of the distance between the centers of the two galaxies, in two different cases. As previously, in the first case, the primary galaxy is active and the values of the parameters are $M_d = 2$ and $M_n = 0.08$, while in the second case the primary galaxy is quiet and therefore, the values of the parameters are $M_d = 2.08$ and $M_n = 0$. The value of the Jacobi integral is always $E_{J2} = -2.6$ and the value of the angular frequency Ω_p is calculated for each particular value of the distance R from equation (3). The range of values regarding the distance between the centers of the two galaxies is again $1.5 \leq R \leq 3$. One can see, in Figure 12, that ⟨L.C.E⟩ decreases almost linearly as the distance R increases when the primary galaxy is active. On the contrary, in the case when the dense and massive nucleus is absent, we observer that the corresponding fitting curve presents a more rapid reduction. Once more, as we have pointed out in Figure 11, for each particular value of the distance R, the average value ⟨L.C.E⟩ is always smaller when the dense and massive nucleus is not present. Moreover, both fitting curves converge to zero, when $R = 3$. Here we must note, that it is well known that the value of the L.C.E is different in each chaotic component (see Saito & Ichimura, 1979). As we have in all cases regular regions and only one unified chaotic area in each $x - p_x$ phase plane, we calculate the average value of the L.C.E by taking 1000 orbits with different and random initial conditions (x_0, p_{x0}) in the chaotic area in each case. Note that, all calculated L.C.Es were different on the fifth decimal point in the same chaotic region.

In Figure 13a-h we present eight representative orbits of 2D dynamical system. Figure 13a shows an quasi-periodic orbit circulating around the primary galaxy, with initial conditions: $x_0 = 0.167$, $y_0 = 0$, $p_{x0} = 0$, while the value of p_{y0} is obtained from the Jacobi integral (13) for all orbits. The value of the Jacobi integral is $E_{J2} = -2.6$ always. The values of all the other parameters are as in Figure 7a. Figure 13b shows a periodic orbit moving around the center of the primary galaxy, with initial conditions: $x_0 = -0.1895$, $y_0 = 0$, $p_{x0} = 0$, while the values of all other parameters are as in Figure 7b. This orbit is characteristic of the 1:2 resonance. In Figure 13c a figure-eight type periodic orbit, characteristic of the 1:3 resonance is shown. The initial conditions are: $x_0 = 0.5788$, $y_0 = 0$, $p_{x0} = 0$, while the value of all other parameters are as in Figure 8a. Figure 13d shows a periodic orbit moving around the primary galaxy with initial conditions: $x_0 = 0.4295$, $y_0 = 0$, $p_{x0} = 0$, while the values of all the other parameters are as in Figure 7b. This orbit is a characteristic example of the 3:3 resonance. In Figure 13e a periodic orbit with initial conditions: $x_0 = 0.7445$, $y_0 = 0$, $p_{x0} = 0$ is presented. This orbit belongs to the family of the 2:3 resonant orbits. The values of all other parameters are as in Figure 7a. In Figure 13f we see a periodic orbit, with initial conditions: $x_0 = -0.9854$, $y_0 = 0$, $p_{x0} = 0$. The values of all other parameters for this orbit are as in Figure 7b. Figure 13g shows a complicated periodic orbit of higher multiplicity, circulating around the primary galaxy, with initial conditions: $x_0 = 0.673$, $y_0 = 0$, $p_{x0} = 0$, while the values of all the other parameters are as in Figure 7a. This orbit produces the set of the nine small islands of the invariant curves in the sticky region, which is embedded in the chaotic area shown in Figure 7a. A chaotic orbit with initial conditions: $x_0 = -0.31$, $y_0 = 0$, $p_{x0} = 0$, is given in Figure 13h. The values of all other parameters for this orbit are as in Figure 7a. We observe that stars moving in

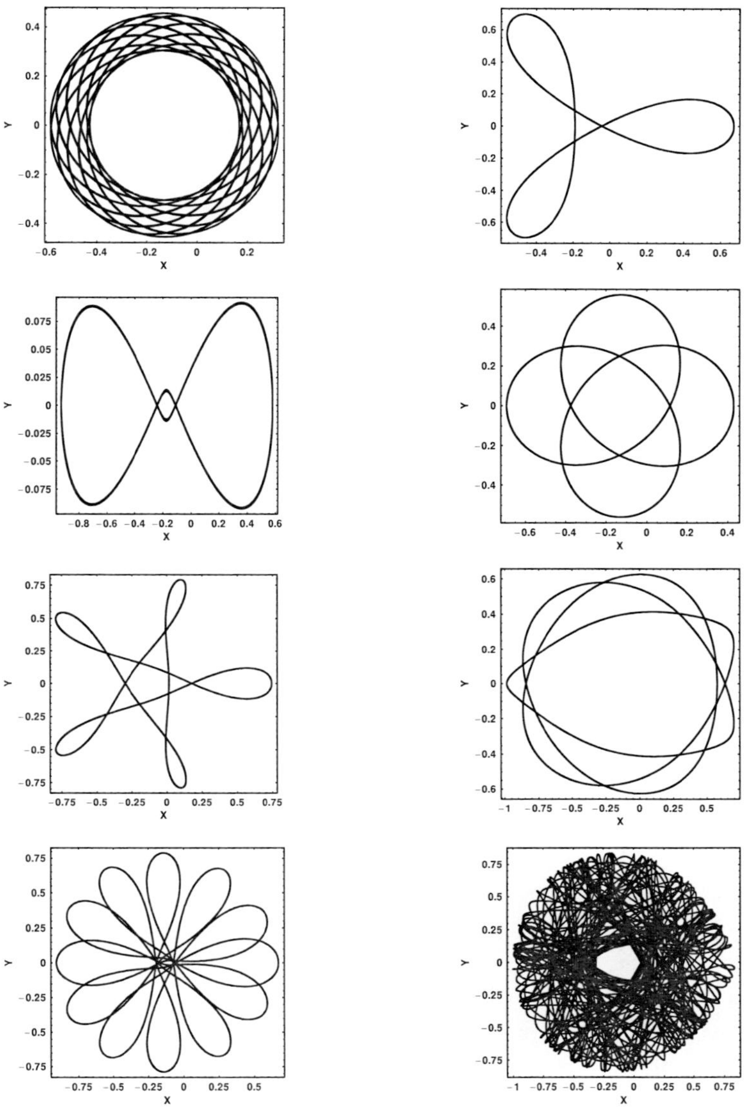

Figure 13. (a-h): Eight representative regular orbits of the 2D dynamical system. The values of the initial conditions and all the other parameters are given in the text.

chaotic orbits pass arbitrarily around the galactic core, without getting close enough. All orbits shown in Figure 13a-h, were calculated for a time period of 150 time units.

Over the last twenty years, dynamical spectra have been frequently used in Galactic Dynamics, as fast and reliable indicators in distinguishing between regular and chaotic motion (see Contopoulos et al., 1995; Contopoulos & Voglis, 1996; Patsis et al., 1997; Caranicolas & Vozikis, 1999; Karanis & Caranicolas, 2001; Caranicolas & Zotos, 2010; Zotos, 2011a,b). In a previous research (see Zotos, 2011a), we have introduced a new dynamical spectrum definition, that is the $S(g)$ spectrum. The main advantage of this new spectrum is that allow us to detect resonant orbits, as it produces as much spectra as the total number of islands of the invariant curves in the corresponding phase plane. The $S(g)$ spectrum was introduced in order to study the nature of orbits in the meridian (r, z) plane, but it can be easily be modified and applied also in the Cartesian (x, y) plane. Therefore, we define the dynamical parameter g as

$$g_i = \frac{x_i + p_{xi} - x_i p_{xi}}{p_{yi}}, \tag{14}$$

where (x_i, p_{xi}, p_{yi}) are the successive values of the (x, p_x, p_y) elements of the 2D orbit, on the Poincaré $x - p_x, y = 0, p_y > 0$ phase plane. We define the new dynamical spectrum of the parameter g, through its distribution function

$$S(g) = \frac{\Delta N(g)}{N \Delta g}, \tag{15}$$

where $\Delta N(g)$ are the number of the parameters g in the interval $(g, g + \Delta g)$, after N iterations. The reader can find more details and theoretical explanations regarding the shapes and behavior of the dynamical spectra, in the pioneer works of Froeschlé et al. (1993), Voglis & Contopoulos (1994) and Contopoulos et al. (1997). By definition, the g parameter is based on a complicate combination of coordinates and momenta. The numerator of the parameter g ia a function of x and p_x while, the denominator is a much more complex function of the same variables, since p_y is related to x and p_x through the Jacobi integral (13). Due to the fact that the values of p_{yi} are the successive values on the Poincaré surface of section, they can never obtain a zero value (always $p_y > 0$), which can infinite the g parameter. There are many reasons for using the $S(g)$ spectrum. Three important reasons are that: (i) it can identify islandic motion of resonant orbits, (ii) it can help us to understand and follow the evolution of sticky motion and (iii) it is much faster than the L.C.E, because it needs about 10^4 time units, while L.C.E needs about 10^5 time units, in order to obtain reliable results.

Figure 14a shows the $S(g)$ spectrum for an orbit producing one of the invariant curves surrounding the central fixed retrograde periodic point shown in the phase plane of Figure 7a. The initial conditions are: $x_0 = 0.47$, $y_0 = 0$, $p_{x0} = 0$, while the value of p_{y0} is always found from the Jacobi integral (13). As expected, one can observe a well defined U-type spectrum, indicating regular motion. The spectrum shown in Figure 14a is symmetrical about the $g = 0$ axis. The fact that in Figure 14a we observe only one U-type spectrum means, that the corresponding orbit is a typical orbit of the 1:1 resonance, forming one invariant curve surrounding the stable retrograde periodic point shown in Figure 7a. The corresponding L.C.E for a time period of 10^5 time units, is given in Figure 14b. One can see that as time evolves, the value of the L.C.E shown in Figure 14b tends asymptotically to zero, which indicates the ordered nature of the orbit and also verifies the result obtained from the $S(g)$ spectrum. In Figure 14c we present the $S(g)$ spectrum for a 2D chaotic orbit. In this case, we see a very complicated, asymmetric spectrum, with a lot of large and abrupt peaks. This orbit belongs to the chaotic sea shown in the phase plane of Figure 7a. The initial conditions are: $x_0 = -0.31$, $y_0 = 0$, $p_{x0} = 0$. The corresponding L.C.E for a time period of 10^5 time units, is given in Figure 14d. We observe, that the pattern of the L.C.E reveals the chaotic nature of the orbit and once more, verifies the fast result given by the $S(g)$ spectrum. Here we must point out that the integration time for the $S(g)$ spectra shown

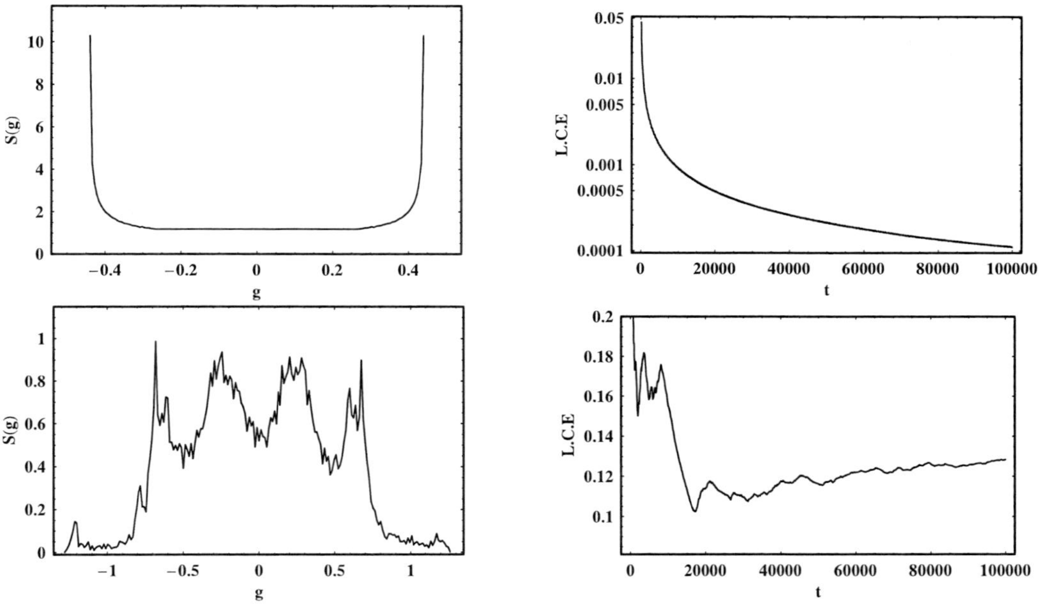

Figure 14. (a-d): (a-upper left): The $S(g)$ spectrum for a regular orbit, (b-upper right): The corresponding L.C.E of the regular orbit, (c-lower left): The $S(g)$ spectrum for a chaotic orbit and (d-lower right): The corresponding L.C.E of the chaotic orbit.

in Figures 14a and c is 5×10^3 time units. Therefore, we can conclude that the $S(g)$ spectrum is a very fast a reliable dynamical indicator.

Our next step, is to deploy the $S(g)$ spectrum in order to study the nature of resonant orbits, which produce multiple islands of invariant curve in the $x - p_x$ phase plane. Figure 15a shows the $S(g)$ spectrum of an orbit producing two islands of invariant curves shown in Figure 7b. The initial conditions of this orbit are: $x_0 = -0.21$, $y_0 = 0$, $p_{x0} = 0$. In Figure 15b we see these two islands of invariant curves inside the limiting curve in the $x - p_x$ phase plane of Figure 7b. Figure 15c shows the $S(g)$ spectrum, for an orbit producing three islands of invariant curves, shown in Figure 7a. The initial conditions are: $x_0 = 0.731$, $y_0 = 0$, $p_{x0} = 0$. Here we observe three U-type spectra, which is as much the number of the islands of invariant curves inside the limiting curve, shown in Figure 15d. Moreover, we observe that the left and the right spectra are exactly symmetrical about the $g = 0$ axis, while the central spectrum lies on both sides of this axis. This indicates, that two of the islands of invariant curves are symmetrical about the x axis, while the third intersects the x axis. In other words, we have the case of a quasi-periodic orbit, with a starting point on the x axis. The integration time for the $S(g)$ spectra shown in Figures 15a and c is 5×10^3 time units. Thus, we came to the conclusion that the $S(g)$ spectrum is very fast and useful dynamical indicator, in order to identify islandic motion of resonant orbits. In this point, we should emphasize that, with some of previously used spectra (see Karanis & Caranicolas, 2002 and references therein), we were able to detect islandic motion, but the number of spectra in some occasions was smaller than the total number of islands of invariant curves, because symmetric islands produce identical spectra. Using the $S(g)$ spectrum and particularly by adding the term $-x_i p_{xi}$ to its definition, we have managed to over through this drawback.

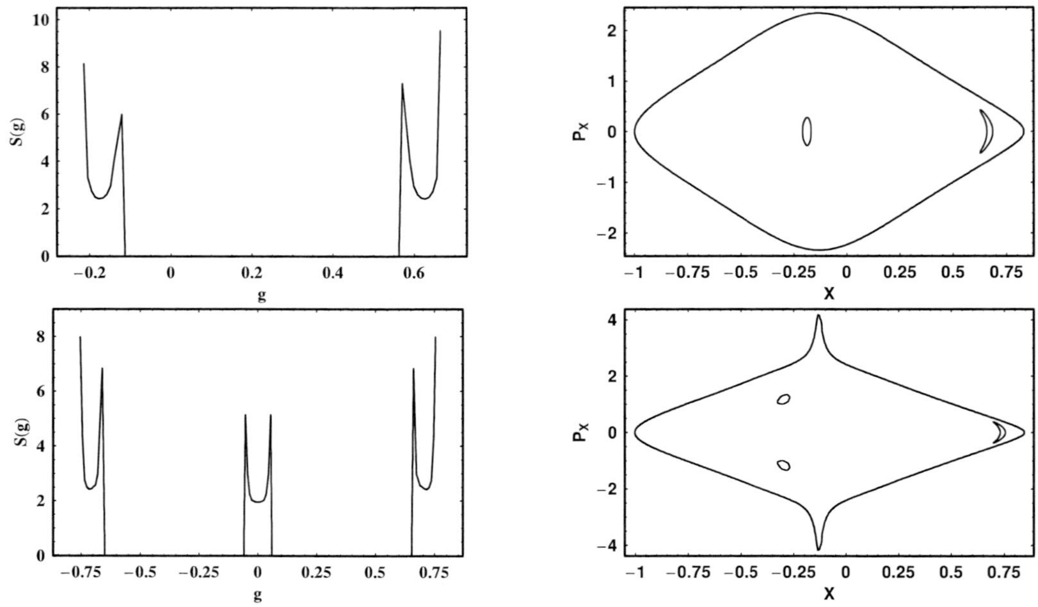

Figure 15. (a-d): (a-upper left): The $S(g)$ spectrum for a resonant orbit producing two islands of invariant curves, (b-upper right): The two islands of invariant curves in the $x-p_x$ phase plane inside the limiting curve, (c-lower left): The $S(g)$ spectrum for a resonant orbit producing three islands of invariant curves and (d-lower right): The three islands of invariant curves in the $x-p_x$ phase plane inside the limiting curve.

4. Structure of the 3D Hamiltonian System

In this Section, we shall investigate the regular or chaotic nature of motion in the 3D Hamiltonian system described by equation (10). In order to keep things simple, we shall use our experience gained from the study of the 2D dynamical system, in order to obtain a clear picture regarding the properties of motion in the 3D dynamical model. We are particularly interested to locate the initial conditions in the 3D dynamical system, producing regular or chaotic orbits. A convenient way to obtain this, is to start from $x-p_x$ phase planes of the 2D system with the same value of the Jacobi integral, used in the 2D system and described in the previous section. Specifically, the regular or chaotic nature of the 3D orbits is found as follows: we choose initial conditions (x_0, p_{x0}, z_0), $y_0 = p_{z0} = 0$, such as (x_0, p_{x0}) is a point on the phase planes of the 2D system. The points (x_0, p_{x0}) lie inside the limiting curve

$$\frac{1}{2}p_x^2 + \Phi_t(x) = E_{J2}, \tag{16}$$

which is the limiting curve containing all the invariant curves of the 2D system. Thus we take $E_J = E_{J2}$. For this purpose a large number of orbits (about 1000) was computed with initial conditions (x_0, p_{x0}, z_0), where (x_0, p_{x0}) is a point in the chaotic regions of the $x-p_x$ phase planes of Figures 7a-b, 8a-b, 9a-b and 10a-b, with all permissible values of z_0 and $p_{z0} = 0$. Remember that, as we are on the phase plane, we have $y_0 = 0$, while in all cases the value of p_{y0} was obtained from the Jacobi integral (10). All tested orbits were found to

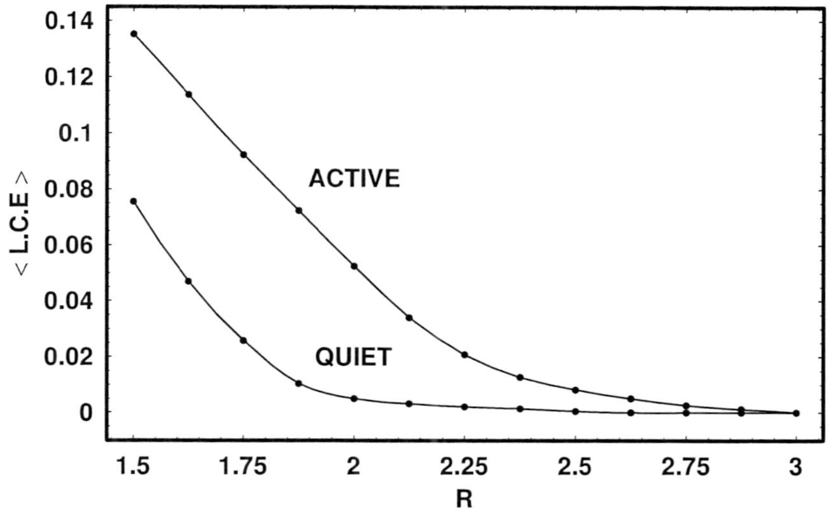

Figure 16. A plot of the average value of the maximum Lyapunov Characteristic Exponent ⟨L.C.E⟩ as a function of the distance R, when the primary galaxy is active or quiet, for the 3D dynamical system.

be chaotic. Therefore, one may conclude that orbits with the initial conditions (x_0, p_{x0}) be a point in the chaotic regions of the 2D phase planes and for all permissible values of z_0, remain chaotic in the 3D system as well.

In Figure 16 we present a plot of the average value of the L.C.E of the 3D system, as a function of the distance between the centers of the two galaxies, in two different cases. As in the previous section, in the first case, the primary galaxy is active and the values of the parameters are $M_d = 2$ and $M_n = 0.08$, while in the second case the primary galaxy is quiet and therefore, the values of the parameters are $M_d = 2.08$ and $M_n = 0$. The value of the Jacobi integral is always $E_J = -2.6$ and the value of the angular frequency Ω_p is calculated for each particular value of the distance R from equation (3). The range of values regarding the distance between the centers of the two galaxies is again $1.5 \leq R \leq 3$. One can see in Figure 16, that the ⟨L.C.E⟩ decreases almost linearly as the distance R increases when the primary galaxy is active, when $R \leq 2.25$. On the contrary, in the case when the dense and massive nucleus is absent, we observer that the corresponding fitting curve presents a more rapid reduction. Once more, as we have pointed out in Figure 12 regarding the 2D system, for each particular value of the distance R, the average value ⟨L.C.E⟩ is always smaller when the dense and massive nucleus is not present. Moreover, both fitting curves converge earlier to zero value, than in Figure 12. The method we use in order to compute the ⟨L.C.E⟩ in each case, is the same as described in Figure 12. For all tested chaotic orbits in the 3D system, the initial value of z_0 is common and equal to 0.1. Here we must point out, that if we compare the plot of the 3D system shown in Figure 16 with that of the 2D system shown in Figure 12, we observe that in each case (active and quiet galaxy respectively) the average values of the L.C.Es of the 3D system are smaller than the corresponding of the 2D system. Therefore, we may conclude that the degree of chaos in the primary galaxy, is smaller when the dynamical system has three degrees of freedom (3D).

Our next step, is to study the character of orbits with initial conditions (x_0, p_{x0}, z_0), $y_0 = p_{z0} = 0$, such as (x_0, p_{x0}) is a point in the regular regions of Figures 7a-b, 8a-b, 9a-b and 10a-b. The phase space of a conservative system of three degrees of freedom has six dimensions, i.e. in Cartesian coordinates $(x, y, z, \dot{x}, \dot{y}, \dot{z})$. For a given value of the Jacobi integral a trajectory lies on a 5-dimensional manifold. In this manifold the surface of section is 4-dimensional. This does not allow us to visualize and interpret directly the structure and the properties of the phase space in dynamical systems of three degrees of freedom. One way to overcome this problem, is to project the surface of section to space with lower dimensions. Later in the same section we will present another alternative. In fact here, we will apply the method introduced by Pfenniger (1984) (see also Revaz & Pfenniger, 2001). We take sections in the plane $y = 0$, $p_y > 0$ of 3D orbits, whose initial conditions differ from the plane parent periodic orbits only by the z component. The set of the resulting 4-dimensional points in (x, p_x, z, p_z) phase space are projected on the (z, p_z) plane. If the projected points lie on a well defined curve, lets call it an "invariant curve", then the motion is regular, while if not, the motion is chaotic. The projected points on the (z, p_z) plane shows nearly invariant curves around the periodic points at $z = 0$, $p_z = 0$, as long as the coupling is weak. When the coupling is stronger the corresponding projections in (z, p_z) plane shows increasing departure of the plane periodic point, up to making a direct orbit a retrograde one and vise versa. Here, we must define what one means by direct and retrograde 3D orbit. If consequents in the (z, p_z) section of the 3D orbit drop in one of the two domains of the corresponding section of 2D orbits at the same value of the Jacobi integral E_J, we can distinguish between direct and retrograde motion. Orbits which visit both domains are intermittently direct or retrograde.

Figure 17a-h shows eight typical (z, p_z) sections of 3D orbits, starting with initial conditions close to different stable periodic points on the (x, p_x) phase planes of the 2D system. In order to obtain the results shown in Figure 17a we have taken the point $(x_0, p_{x0}) = (0.25, 0)$ representing approximately the position of the retrograde periodic point in the (x, p_x) phase plane of Figure 7a. Similarly in Figure 17b we can observe the (z, p_z) projections near the retrograde periodic point in the (x, p_x) phase plane of Figure 7b. The position of the periodic point is $(x_0, p_{x0}) = (0.25, 0)$. Moreover, in Figure 17c we see the (z, p_z) projections near the retrograde periodic point of active galaxy, shown in Figure 8a. The position of the periodic point is $(x_0, p_{x0}) = (0.2, 0)$. One more example is given in Figure 17d, where one can see the (z, p_z) projections near the direct periodic point of quiet galaxy, shown in Figure 8b. The position of the periodic point this time is $(x_0, p_{x0}) = (-0.7, 0)$. In order to obtain the results shown in Figure 17e we have taken the point $(x_0, p_{x0}) = (-0.72, 0)$ representing approximately the position of the direct periodic point in the (x, p_x) phase plane of Figure 9a. Similarly in Figure 17f we can observe the (z, p_z) projections near the retrograde periodic point in the (x, p_x) phase plane of Figure 9a. The position of the periodic point is $(x_0, p_{x0}) = (0.18, 0)$. Furthermore, in Figure 17g we see the (z, p_z) projections near the direct periodic point of active galaxy, shown in Figure 10a. The position of this periodic point is $(x_0, p_{x0}) = (-0.75, 0)$. Finally, in Figure 17h, one can observe the (z, p_z) projections near the direct periodic point of quiet galaxy, shown in Figure 10b. The position of this periodic point is $(x_0, p_{x0}) = (-0.74, 0)$. Note that in all cases the numerical results indicate, that for small values of z_0 the motion is regular, while for larger values of z_0 the motion become chaotic. The integration time for each "invariant curve" shown in Figure 17a-h, is 3000 time

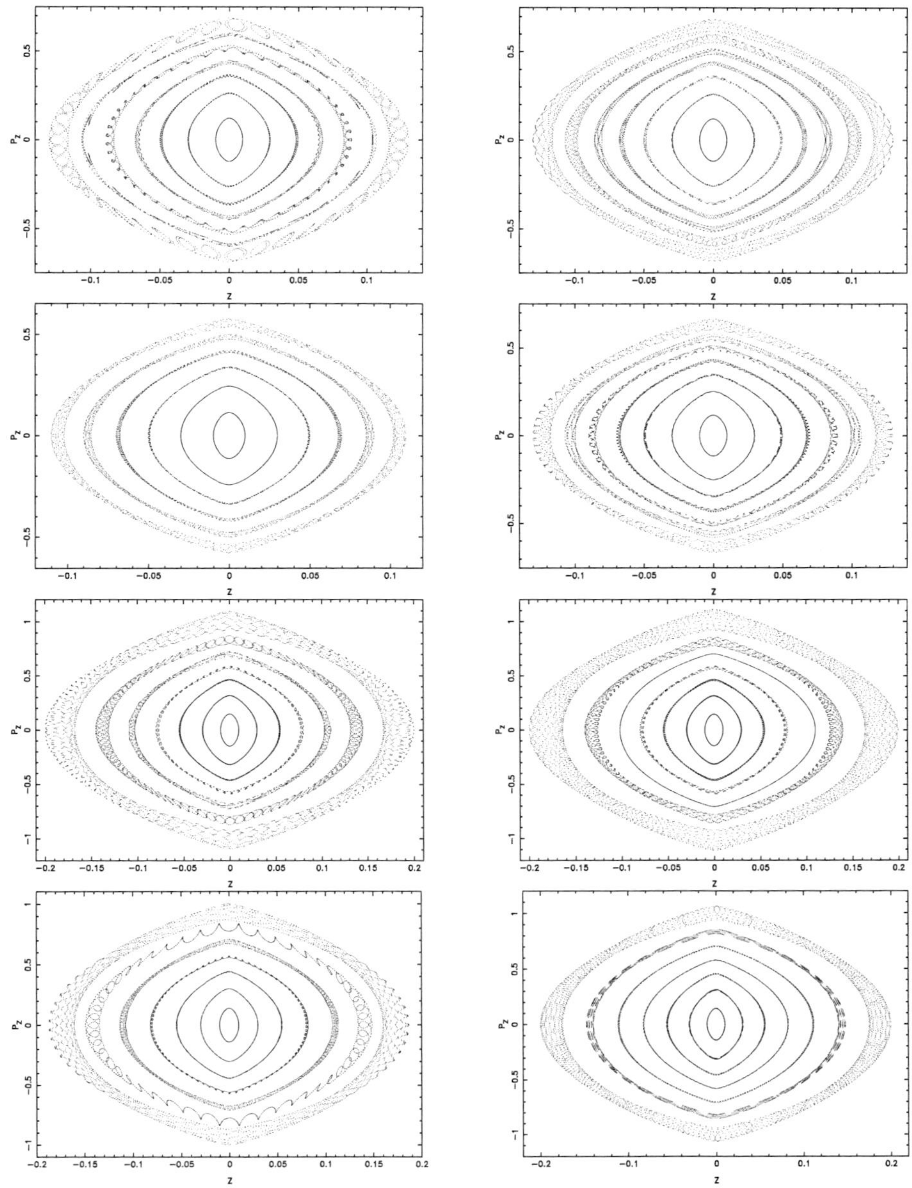

Figure 17. (a-h): Projections of the sections of 3D orbits with the plane $y = 0$ when $p_y > 0$. The set of the four dimensional points (x, p_x, z, p_z) is projected on the $z - p_z$ plane.

units, while the value of the Jacobi integral is always $E_J = -2.6$. Numerical calculations not given here, suggest that the above method can be applied in all regular regions around each retrograde or direct stable periodic point of Figures 7a-b, 8a-b, 9a-b and 10a-b.

We must emphasize that the results shown in Figure 17a-h are rather qualitative and can be considered as an indication that the transition from regularity to chaos in 3D orbits, occurs as the value of z_0 increases. In order to form a more complete and accurate view of the phase space in the 3D system, we computed a large number of 3D orbits (approximately

1000) near in each periodic point of the (x, p_x) phase planes of 2D system, for different initial conditions (x_0, p_{x0}) and also for different values of z_0. Our target was to determine the average minimum value of z_0 for which the nature of a 3D orbit changes from regular to chaotic. Table 1 shows the value $\langle z_{min} \rangle$ near the direct and retrograde stable periodic points of Figures 7a-b, 8a-b, 9a-b and 10a-b, for four different values of distance between the centers of the two galaxies R. From Table 1 we can induce three important results: (i) In every case (Active or Quiet primary galaxy) the minimum value of z_0 in the regions near the retrograde stable periodic points, is always larger than the corresponding of the regions of the direct periodic points. (ii) If we compare each kind of galaxy regarding its nuclear activity, (Active or with Quiet) for the same distance R, we observe that when the dense and massive nucleus is not present at the galactic core (Quiet - non Active primary galaxy), the 3D orbits can approach higher values of z_0 and remain regular. On the other hand, when the dense and massive nucleus is present, the value of $\langle z_{min} \rangle$ is smaller. (iii) As the distance between the centers of the galaxies increases, the minimum initial value of z_0 for which a 3D can remain regular, increases in both cases (Active or Quiet primary galaxy). This means that as the two galaxies are in large distances, their mutual interactions are weak enough and therefore, the majority of the 3D orbits are ordered. Moreover, in large distances one can conclude, that the main responsible for the observed chaotic motion is the nuclear activity of the dense and massive nucleus in the core of the primary galaxy.

Table 1. Average value of the minimum z_0 near the direct and retrograde periodic points, for four different values of the distance R.

Distance	Case	Region	$\langle z_{min} \rangle$
$R = 1.5$	Active	Direct	-
		Retrograde	0.0792
	Quiet	Direct	-
		Retrograde	0.0802
$R = 2.0$	Active	Direct	0.0865
		Retrograde	0.0912
	Quiet	Direct	0.0896
		Retrograde	0.0964
$R = 2.5$	Active	Direct	0.0983
		Retrograde	0.1025
	Quiet	Direct	0.1012
		Retrograde	0.1107
$R = 3.0$	Active	Direct	0.1098
		Retrograde	0.1157
	Quiet	Direct	0.1185
		Retrograde	0.1214

So far, we have seen that 3D orbits with initial conditions (x_0, p_{x0}, z_0), such as (x_0, p_{x0}) is a point in the chaotic regions of the 2D system, for all permissible values of z_0 are chaotic. On the other hand, the nature (ordered or chaotic) of 3D orbits with initial conditions (x_0, p_{x0}, z_0), such as (x_0, p_{x0}) is a point in the regular regions around the stable direct or retrograde periodic points of the 2D system, depends on the particular value of z_0, as we can see in Table 1. We did not feel that it was necessary to try to define the values of $\langle z_{min} \rangle$ for each regular region of the 2D system corresponding to secondary resonances which are represented by multiple small islands of invariant curves in the $x - p_x$ phase planes, shown in Figures 7a-b, 8a-b, 9a-b and 10a-b. Numerical results indicate, that 3D orbits with initial conditions (x_0, p_{x0}, z_0), such as (x_0, p_{x0}) is a point in the regular regions corresponding to secondary resonances of the 2D system, remain regular for $\langle z_{min} \rangle \lesssim 0.052$, while for larger values of z_0, they change their character from regular to chaotic.

In Figure 18a-h we present eight typical orbits of the 3D dynamical system. Figure 18a shows a regular quasi-periodic orbit circulating around the primary galaxy, with initial conditions: $x_0 = 0.167, y_0 = 0, z_0 = 0.01, p_{x0} = p_{z0} = 0$, while the value of p_{y0} is always found from the Jacobi integral (10). The value of the Jacobi integral is always $E_J = -2.6$. In Figure 18b we observe a 3D periodic orbit characteristic of the 1:2 resonance. This orbit has initial conditions: $x_0 = -0.1895, y_0 = 0, z_0 = 0.02, p_{x0} = p_{z0} = 0$. Figure 18c shows a 3D periodic orbit with initial conditions: $x_0 = 0.5788, y_0 = 0, z_0 = 0.01, p_{x0} = p_{z0} = 0$, circulating around the center of the primary galaxy. In Figure 18d we see quasi-periodic orbit, with initial conditions: $x_0 = 0.4295, y_0 = 0, z_0 = 0.015, p_{x0} = p_{z0} = 0$. A 3D resonant orbit with initial conditions: $x_0 = 0.7445, y_0 = 0, z_0 = 0.01, p_{x0} = p_{z0} = 0$, moving around the galactic center, is shown in Figure 18e. In Figure 18f we observe a 3D periodic orbit circulating around the primary galaxy. This orbit has initial conditions: $x_0 = -0.9854, y_0 = 0, z_0 = 0.01, p_{x0} = p_{z0} = 0$. In Figure 18g we observe a complicated 3D resonant periodic orbit with initial conditions: $x_0 = 0.673, y_0 = 0, z_0 = 0.005, p_{x0} = p_{z0} = 0$. Figure 18h shows a 3D chaotic orbit with initial conditions: $x_0 = -0.31, y_0 = 0, z_0 = 0.1, p_{x0} = p_{z0} = 0$. This orbit goes arbitrarily close to the primary galaxy and it is deflected to higher values of z, on approaching the dense and massive nucleus. We must note, that in all 3D orbits shown in Figure 18a-h the initial conditions (x_0, p_{x0}) and the values of all the other parameters (M_d, M_n, R), are as in the corresponding 2D orbits shown in Figure 13a-h. Moreover, we observe that all regular 3D orbits shown in Figure 18a-g, stay neat to the galactic plane and therefore, support the disk structure of the primary galaxy. The numerical integration time for all 3D orbits shown in Figure 18a-h, is 200 time units.

In the previous section, we have used the $S(g)$ dynamical spectrum, in order to study the nature of the orbits in the 2D system. The $S(g)$ spectrum can also be used for the study of the character of the 3D orbits. Note, that the coupling of the third component, z, carrying all the information about the 3D motion, is hidden in the definition of the $S(g)$ spectrum, but in any case it affects the values of the elements of the orbit $\left(x, p_x, p_y\right)$ entering relation (14). In an attempt to construct a new dynamical spectrum suitable especially for 3D orbits, we have introduced and and used in Zotos (2011b) the $S(w)$ spectrum. Here we must remind to the reader that this spectrum is the distribution function of the w parameter

$$S(w) = \frac{\Delta N(w)}{N \Delta w}, \qquad (17)$$

where $\Delta N(w)$ are the number of the parameters w in the interval $(w, w + \Delta w)$, after N itera-

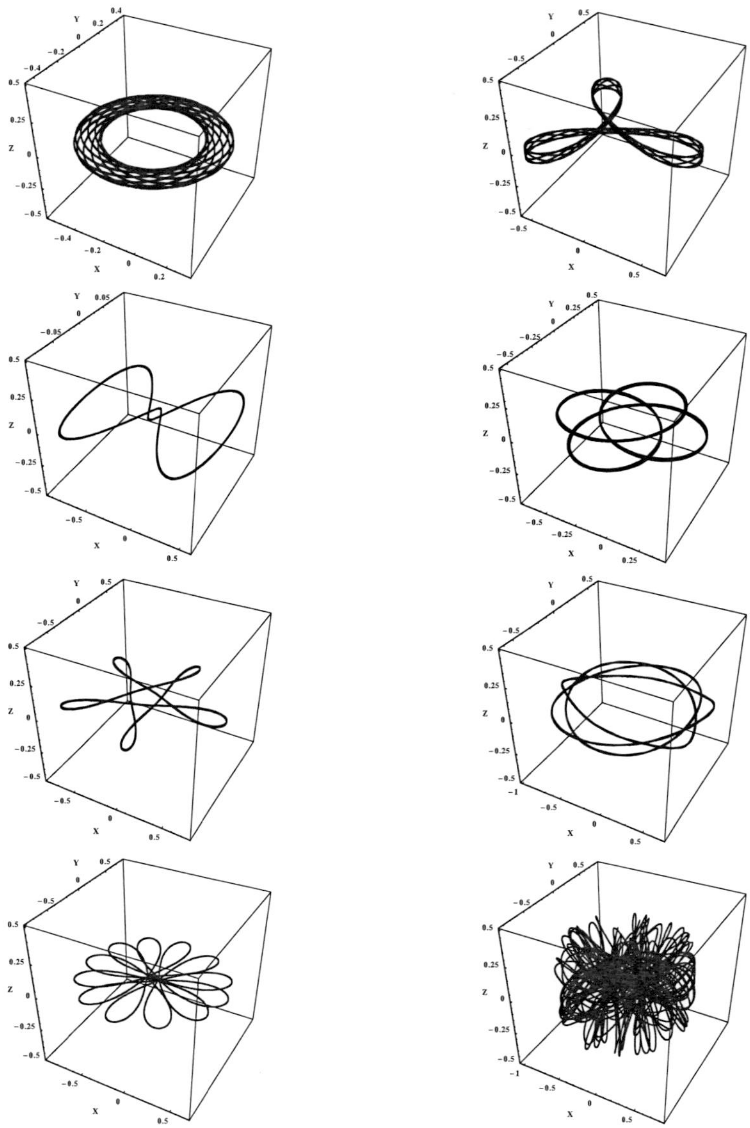

Figure 18. (a-h): Eight representative regular orbits of the 3D dynamical system. The initial conditions and more details regarding the values of all the other parameters are given in the text.

tions. The parameter w, is defined as

$$w_i = \frac{(x_i - p_{xi}) - (z_i - p_{zi})}{p_{yi}}, \tag{18}$$

where $\left(x_i, z_i, p_{xi}, p_{yi}, p_{zi}\right)$ are the successive values of the elements $\left(x, z, p_x, p_y, p_z\right)$ of the 3D orbit. By definition, the w parameter is based on a complicate combination of coordinates and velocities of the 3D orbit. There are several reasons, for introducing the $S(w)$ spectrum. Three important reasons are that: (i) It can identify 3D islandic motion as is produces as many U-type structures as the total number of the 3D invariant tori. (ii) It can help us to understand the evolution of 3D sticky orbits and (iii) It is much faster than other dynamical indicators, such as the L.C.E, S.AL.I and G.AL.I. All these indicators need integration time periods of order of 10^5 time units, in order to give reliable results, while the $S(w)$ spectrum can provide us with reliable evidence regarding the nature of a 3D orbit after only 10^3 time units.

Let us now explain another interesting method through which we can reveal the properties of motion in Hamiltonian systems with three degrees of freedom. The structure of the 4D space phase space was examined for the first time in the pioneer work of Froeschlé (1970, 1972). In that work he used stereoscopic views and the method of slices in order to understand the structure of tori, that appeared at the neighborhood of stable periodic orbits. Similar methods have been applied by Martinet & Magnenat (1981), Contopoulos et al. (1982) and Magnenat (1982) for studying the 3D projections of invariant tori in the 4D surface of section or in the phase space of a 4D symplectic map. The 2D projections of such invariant tori have been examined on various 2D planes in detail (e.g. Skokos, Contopoulos & Polymilis 1997, 1999). In the present work, we shall use the method proposed by Patsis & Zachilas (1994). In this method, we first define the surface of section that we will use, e.g. $y = 0$ with $p_y > 0$. Then, we select a 3D subspace projection e.g. (x, p_x, z) of the (x, p_x, z, p_z) 4D phase space. If the 3D projection is a well formed 3D invariant torus then the initial conditions correspond to a regular orbit. On the other hand, if instead of a well defined torus, we find a vague cloudy structure in the (x, p_x, z) projection, then the corresponding 3D orbit is chaotic.

Figure 19a depicts a well formed 3D invariant torus, with initial conditions: $x_0 = 0.54$, $y_0 = 0$, $z_0 = 0.01$, $p_{x0} = p_{z0} = 0$, while the value of p_{y0} is always found from the Jacobi integral (10). The value of the Jacobi integral is always $E_J = -2.6$. The values of all the other parameters are as in Figure 7a. It is evident that this orbit is a regular 1:1 resonant orbit. In Figure 19b we see the $S(w)$ spectrum for this orbit. As expected, we observe a well defined U-type spectrum corresponding to regular motion. Figure 19c shows a pair of well formed 3D invariant tori. The initial conditions for this orbit are: $x_0 = -0.21$, $y_0 = 0$, $z_0 = 0.0025$, $p_{x0} = p_{z0} = 0$. The values of all the other parameters are as in Figure 7b. The presence of the two 3D invariant tori, indicates that these initial conditions correspond to a 1:2 resonant 3D orbit. This fact can be justified by looking at Figure 19d, where the $S(w)$ spectrum of this orbit is presented. Here, one can see that the number of the U-type spectra is equal to the number of the 3D invariant tori. A similar example is shown in Figure 19e. In this case we see a triplet of 3D invariant tori. The initial conditions for this orbit are: $x_0 = 0.731$, $y_0 = 0$, $z_0 = 0.0025$, $p_{x0} = p_{z0} = 0$, while the values of all the other parameters are as in Figure 7a. The number of the 3D invariant tori, reveals that this orbit belongs to the family of the 2:3 resonant orbits. This result coincides with the result obtained from the $S(w)$ spectrum and it is shown in Figure 19f. Once more, the number of the U-type spectra is equal to the total number of the 3D invariant tori. Things are quite different in the (x, p_x, z) projection shown in Figure 19g. Here, instead of invariant tori, we observe a vague cloudy structure of random points. The initial conditions for this orbit are: $x_0 = -0.31$, $y_0 = 0$, $z_0 = 0.1$, $p_{x0} = p_{z0} = 0$, while the values of all the other parameters are as in Figure 7a. This cloudy structure indicates that this is a 3D chaotic orbit. We can conclude to the same result from the Figure 19h, where the $S(w)$ spectrum of the 3D chaotic orbit is presented. We see

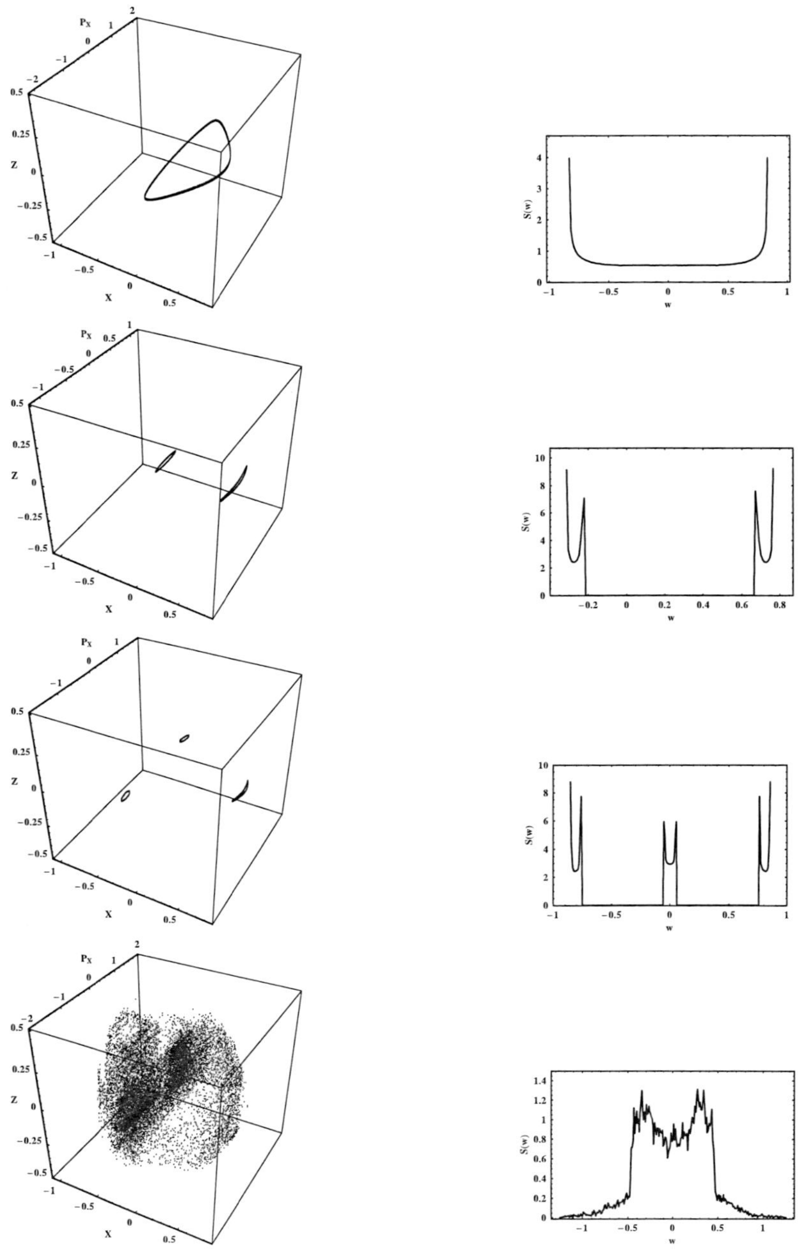

Figure 19. (a-h): The (x, p_x, z) 3D projections of the (x, p_x, z, p_z) 4D phase space and the corresponding $S(w)$ spectra for the 3D orbits.

a complicated, asymmetric structure with a lot of large and abrupt peaks in its pattern. This strongly suggests the chaotic nature of the orbit. The integration time for the $S(g)$ spectra shown in Figures 19b, d, f and h is 5×10^3 time units, while for the 3D (x, p_x, z) projections shown in Figures a, c, e and g is 2×10^4 time units. Thus, we came to the conclusion that the $S(w)$ spectrum is very fast and reliable dynamical indicator, in order to identify islandic

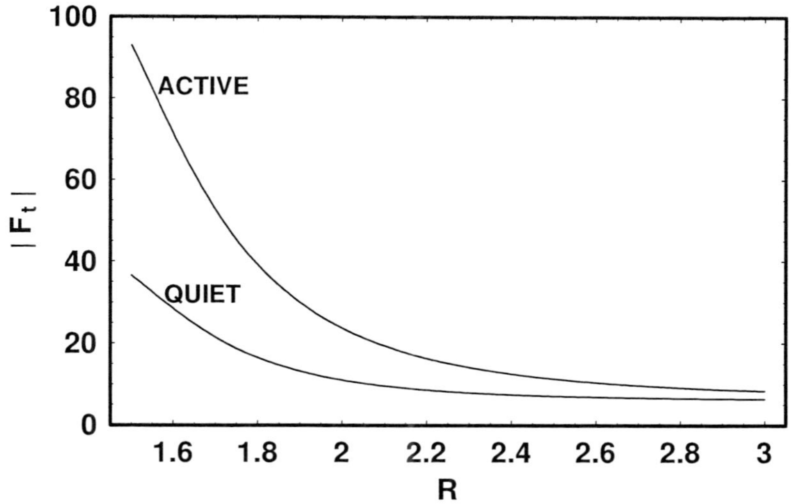

Figure 20. A plot of the total $|F_t|$ force as a function of the distance R, when the primary galaxy is active or quiet.

motion of 3D resonant orbits.

5. A Theoretical Approach

In the present section, we shall present some theoretical arguments, together with elementary numerical calculations, in order to explain the numerically obtained relationships, given in Figures 11, 12 and 16. Moreover, we will quote different kinds of theoretical techniques in an attempt to explain and study deeper the dynamical structure of the 2D or 3D Hamiltonian system.

Since the potential of the nucleus is integrable with spherical symmetry, the the observed chaotic phenomena in our dynamical system, should derive from the total $|F_t|$ force. Figure 20 shows a plot of the $|F_t|$ force as a function of the distance R between the two galaxies at the point $P_c(x_0, y_0, z_0) = (-0.12, 0.02, 0.01)$. This point is very close to the center of the primary galaxy, where the chaotic phenomena are more prominent. The value of the Jacobi in integral is $E_J = -2.6$. One may observe, from the plot shown in Figure 20 that the two curves correspond to two different cases (active and quiet primary galaxy) have almost the same pattern. In both cases, the $|F_t|$ force decreases, tending asymptotically to zero value, as the value of the distance R increases. Here, we must point out that the values of the $|F_t|$ force are larger when the primary galaxy is active and therefore, the corresponding curve is above the one corresponds to the quiet galaxy. Thus, we conclude that when a massive and dense nucleus is present in the center of the primary galaxy the extent of the chaotic orbits is larger, than in the case in which the primary galaxy is quiet. One can see that, the pattern between the $|F_t|$ force and the distance R (shown in Figure 20), is very similar to those between the chaotic percentage $A\%$ or the L.C.E (shown in Figures 11, 12 and 16 respectively), which have been obtained numerically.

Furthermore, it would be of particular interest to study the structure of the velocity

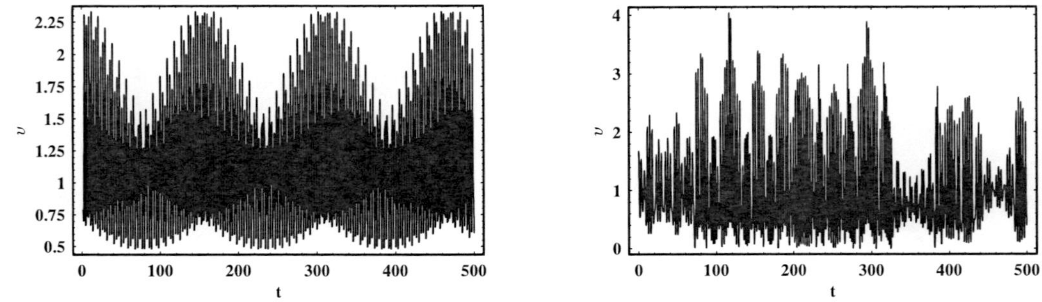

Figure 21. (a-b): (a-left): The total velocity profile of the 3D orbit shown in Figure 18a. We observe a nearly periodic pattern and (b-right): The total velocity profile for the chaotic 3D orbit shown in Figure 18g. In this case there are abrupt changes in the profile's pattern indicating chaotic motion.

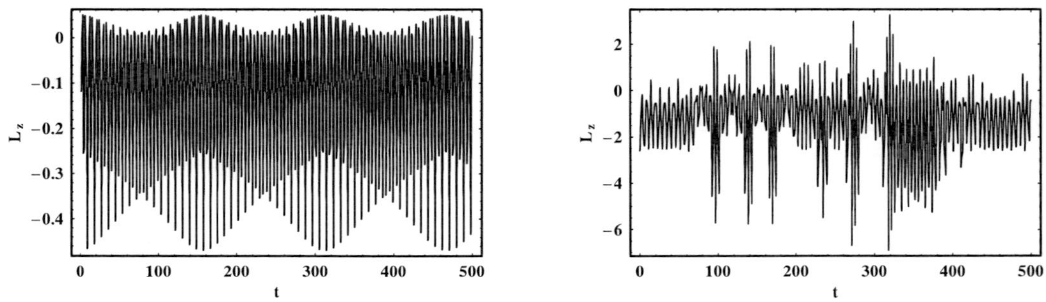

Figure 22. (a-b): A plot of L_z component of the total angular momentum vs. time for (a-left): the regular 3D orbit shown in Figure 18a and (b-right): the chaotic 3D orbit shown in Figure 18h.

profile, that is the plot of the total velocity of a test particle (star), $\upsilon = \sqrt{\dot{x}^2 + \dot{y}^2 + \dot{z}^2}$ as a function of time, for ordered and chaotic motion. Figure 21a shows the velocity profile for a time period of 500 time units corresponding to the regular 3D orbit shown in Figure 18a. Here, the velocity profile is quasi-periodic and the maximum value of the velocity is about 450 km/s. This suggest that regular 3D motion is made in small velocities. Figure 21b shows the velocity profile for a time period of 500 time units corresponding to the chaotic 3D orbit shown in Figure 18h. In this case we can discuss two aspects. The first is that the velocity obtains high values up to 790 km/s and second is that the velocity profile appears to be highly asymmetric displaying abrupt changes and large deviations between the maxima and also between the minima in the $[\upsilon - t]$ plot. Therefore, we can conclude that the chaotic 3D motion is made in high and abruptly changing velocities.

The physical parameter playing an important role on the orbital behavior of the stars, is the L_z component of the total angular momentum. From our previous experience, we know that low angular momentum stars on approaching a dense and massive nucleus are scattered off the galactic plane, displaying chaotic motion (Caranicolas & Innanen, 1991; Caranicolas & Papadopoulos, 2003, Caranicolas & Zotos, 2010; Zotos, 2011a,c). Of course here in 3D phase space things are more complicated than in axially symmetric dynamical

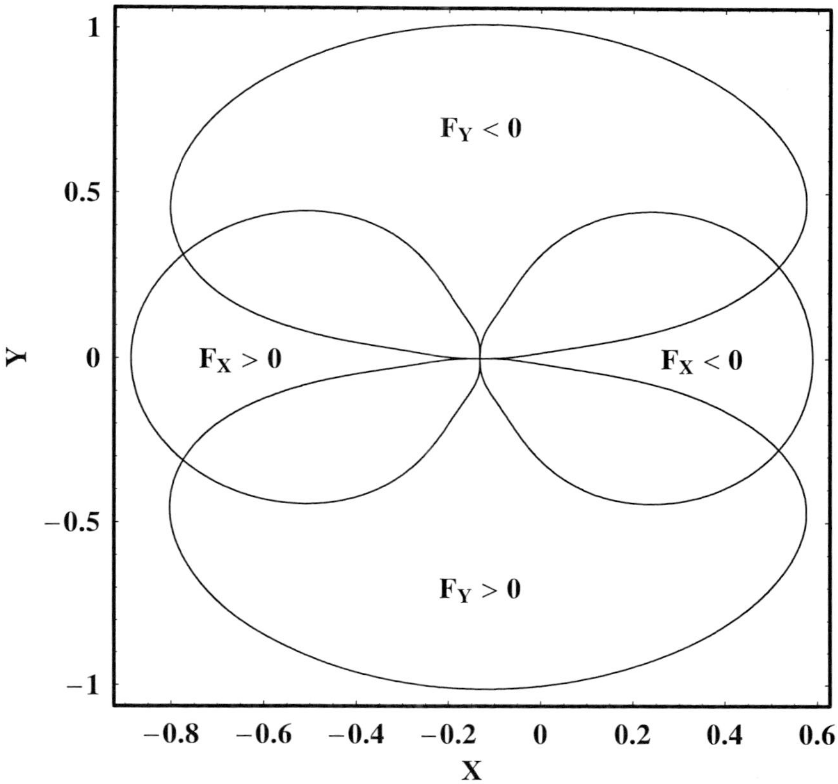

Figure 23. Contours of the F_x = *const* (elliptic shaped) together with the contours F_y = *const* (figure eight shaped). Details are given in the text.

models, where the L_z component was conserved. As the motion takes place in a rotating non axially symmetric system, the L_z component, which is given by

$$L_z = x\dot{y} - \dot{x}y - \Omega_p \left(x^2 + y^2\right), \tag{19}$$

is not conserved. Nevertheless, we can compute numerically its mean value $\langle L_z \rangle$ using the formula

$$\langle L_z \rangle = \sum_{i=0}^{n} L_{zi} . \tag{20}$$

Our numerical calculations suggest that the chaotic orbits have low values of $\langle L_z \rangle$, while regular orbits obtain high values of $\langle L_z \rangle$. Figure 22a shows a plot of the evolution of L_z component with the time for the regular orbit of Figure 18a. In this case we observe that L_z is nearly a quasi-periodic function of time, while $\langle L_z \rangle = -0.12271$. Figure 22b is similar to 22a but the chaotic orbit shown in Figure 18h. Here one can see abrupt changes of the L_z during the chaotic motion, while for this chaotic orbit we have $\langle L_z \rangle = -1.24662$. In both cases the time interval of the numerical integration is 500 time units, while $n = 10^4$.

In what follows we shall come to present a semi-theoretical analysis, in order to give a more detailed picture of the structure of 2D dynamical system and its behavior. The forces

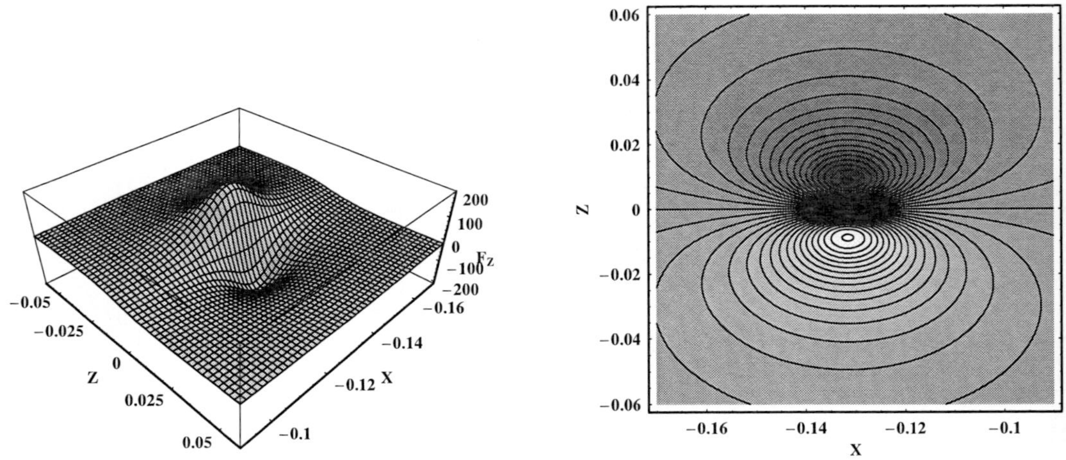

Figure 24. (a-b): (a-left): A 3D plot of the value of the F_z force on the (x, z) plane and (b-right): Contours of the projections $F_z = const$ on the (x, z) plane. Lighter colors indicate higher values of F_z. For positive values of z the F_z force is negative, while for negative values of z the F_z force is positive.

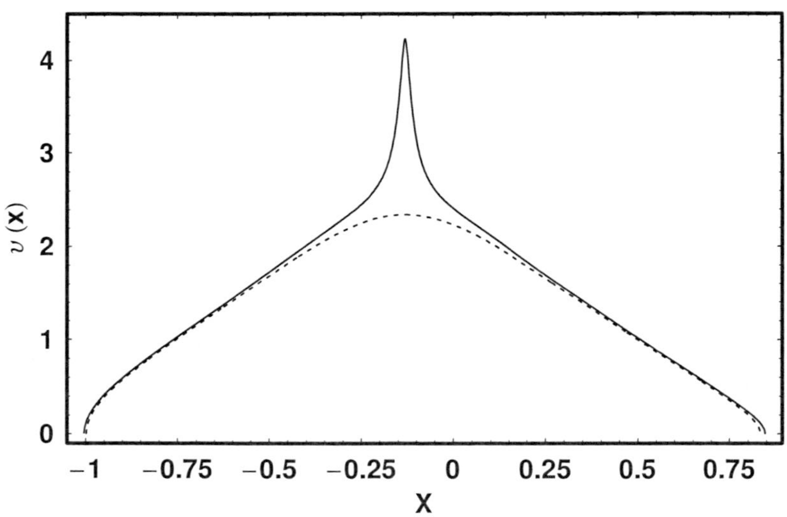

Figure 25. A plot of the total velocity $\upsilon(x)$ as a function of the distance x, for the two cases (Active and Quiet primary galaxy).

acting on a test particle along the x and y axis are given by the equations

$$\begin{aligned} F_x &= -\frac{M_n\,(x - x_1)}{\left(r_{a1}^2 + c_n^2\right)^{3/2}} - \frac{M_d\,(x - x_1)}{\left[b^2 + r_{a1}^2 + (a + h)^2\right]^{3/2}} - \frac{M_s\,(x - x_2)}{r_{a2}^3} + \Omega_p^2 x\,, \\ F_y &= -\frac{M_n y}{\left(r_{a1}^2 + c_n^2\right)^{3/2}} - \frac{M_d y}{\left[b^2 + r_{a1}^2 + (a + h)^2\right]^{3/2}} - \frac{M_s y}{r_{a2}^3} + \Omega_p^2 y\,. \end{aligned} \tag{21}$$

It is obvious from equations (21) that the strength of both forces increases as the mass of the nucleus or the disk increases or their scale lengths decrease. Figure 23 shows the contours of $F_x = const$ together with the contours $F_y = const$. The values of all other parameters are as in Figure 2. The contours of $F_x = const$ look like ellipses, while the contours of $F_y = const$ have a figure-eight shape. Inside the first "ellipse" we have $F_x > 0$, while inside the second "ellipse" we have $F_x < 0$. On the other hand, inside the lower figure-eight curve we have $F_y > 0$, while inside the upper figure-eight curve we have $F_y < 0$. Looking carefully near the galactic center, we observe that there are areas where F_x is positive and F_y is negative at the same time and vise versa. There are also areas, where both forces are positive or negative at the same time. Therefore, we conclude that near the center of the primary galaxy, there are strong attractive and repulsive forces acting on the test particle (star). These forces are responsible for the chaotic scattering of the star near the galactic center, leading to chaotic motion.

At this point, we can expand the semi-theoretical analysis described above, in order to give a more detailed and complete picture of the structure of 3D dynamical system and its behavior. The force acting on a test particle along the z axis is given by the equation

$$F_z = -\frac{M_n z}{\left(r_1^2 + c_n^2\right)^{3/2}} - \frac{M_d\,(a + r_z)\,z}{r_z\left[b^2 + r_{a1}^2 + (a + r_z)^2\right]^{3/2}} - \frac{M_s z}{r_2^3}, \tag{22}$$

where $r_z = \sqrt{h^2 + z^2}$. It is obvious from equation (22) that the strength of the F_z force increases as the mass of the nucleus or the disk increases or their scale lengths decrease. Figure 24a shows a 3D plot of the value of the F_z force on the (x, z) plane. Figure 24b depicts the contours of the projections $F_z = const$ on the (x, z) plane. We observe, that for positive values of z the F_z force is negative, while for negative values of z the value of the F_z is positive, near the center of the primary galaxy. Therefore, lighter colors on the lower half part of the (x, z) plane indicate higher values of the F_z force, while darker colors on the upper half part of the same plane indicate lower values of the F_z force. One can observe, that near the active galactic core which host a dense and massive nucleus, the test particle experiences a very strong F_z force. The values of all the parameters in Figure 24a-b, are as in Figure 2.

Last but not least, it would be also interesting to investigate the behavior of the total velocity near the galactic core, as a function of the distance x. In order to do that, we consider the limiting curve, that is, the curve containing all the invariant curves on the $x - p_x$ phase plane. This can be obtained, if we set $y = z = p_y = p_z = 0$ in equation (10) yielding to

$$\frac{1}{2}p_x^2 = \frac{1}{2}\upsilon^2 = E_J - \Phi_t(x), \tag{23}$$

where we have set $p_x = \upsilon$ because at the limiting curve the p_x velocity is the total velocity (see also Papadopoulos & Caranicolas, 2006). In Figure 25, we can observe two plots of the total velocity υ as a function of the distance x, derived using relation (23). The values of all the parameters are as in Figure 2. The dashed line corresponds to the case of the quiet primary galaxy, while the solid line corresponds to the case when the primary galaxy host a dense and massive nucleus in its core and therefore, is active. We see that the velocity is about the same for all values of x, except near the galactic center, that is when $x \simeq -0.13$, where higher velocities correspond to active galactic center, hosting massive and dense nucleus.

Table 2. Radii of Lindblad resonances when $R = 2.35$ and $\Omega_p = 0.419146$, for both active and quiet primary galaxy.

Case	Resonance	Region	r_1	r_2
Active	$\Omega_p = \Omega - 2\kappa/3$	Direct	0.0149	-
		Retrograde	0.0142	0.2385
	$\Omega_p = \Omega - 3\kappa/4$	Direct	0.0246	-
		Retrograde	0.0217	0.3548
	$\Omega_p = \Omega - 5\kappa/8$	Direct	0.0127	0.3886
		Retrograde	0.0122	-
	$\Omega_p = \Omega - 7\kappa/9$	Direct	0.0257	-
		Retrograde	0.0238	0.3912
Quiet	$\Omega_p = \Omega - 2\kappa/5$	Direct	0.0123	0.5527
		Retrograde	0.0114	0.1342
	$\Omega_p = \Omega - 3\kappa/4$	Direct	0.0120	-
		Retrograde	0.0119	0.2351
	$\Omega_p = \Omega - 3\kappa/7$	Direct	0.0095	0.6782
		Retrograde	0.0091	-
	$\Omega_p = \Omega - 4\kappa/9$	Direct	0.0207	-
		Retrograde	0.0196	0.2608

One of the factors responsible for the chaotic motion and other resonance phenomena, such as islandic motion corresponding to multiple islands of invariant curves or invariant tori, are several inner Lindblad resonances

$$\Omega_p = \Omega - \frac{n}{m}\kappa, \tag{24}$$

where Ω and κ indicate the circular and the epicycle frequency of the star respectively, while m and n are integers. The main resonances for the two cases (Active and Quiet primary galaxy) together with the corresponding resonance radii r_1 and r_2, when $R = 2.35$ and $\Omega_p = 0.419146$ are given in Table 2.

Figure 26a shows a plot of the curves $\Omega - n\kappa/m$ in the case where the primary galaxy is active, as a function of the radius r, when $R = 2.35$. We choose this particular value of the distance between the two galaxies, because at this distance we can observe a large variety of Lindblad resonances. The numbers 1, 2, 3 and 4 indicate the curves: $\Omega - 2\kappa/3$, $\Omega - 3\kappa/4$, $\Omega - 5\kappa/8$ and $\Omega - 7\kappa/9$ respectively. The straight lines are the curves $\Omega_p = \pm 0.419146$. here, the numbers 1, 2, 3 and 4 indicate the curves: $\Omega - 2\kappa/5$, $\Omega - 3\kappa/4$, $\Omega - 3\kappa/7$ and $\Omega - 4\kappa/9$ respectively. The values of all the other parameters are as in Figure 2. Figure 26b is similar to Figure 26a but when the primary galaxy is quiet. As we can see there is

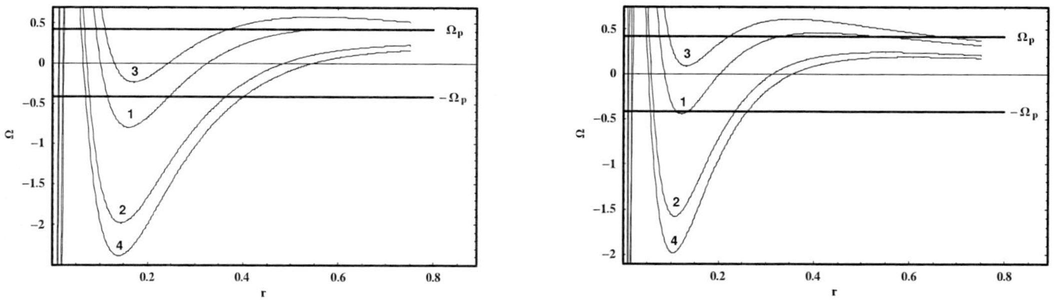

Figure 26. (a-b): The curves $\Omega - n\kappa/m$ vs r, when $R = 2.35$ and $\Omega_p = 0.419146$, when (a-left): the primary galaxy is active and (b-right): the primary galaxy is quiet.

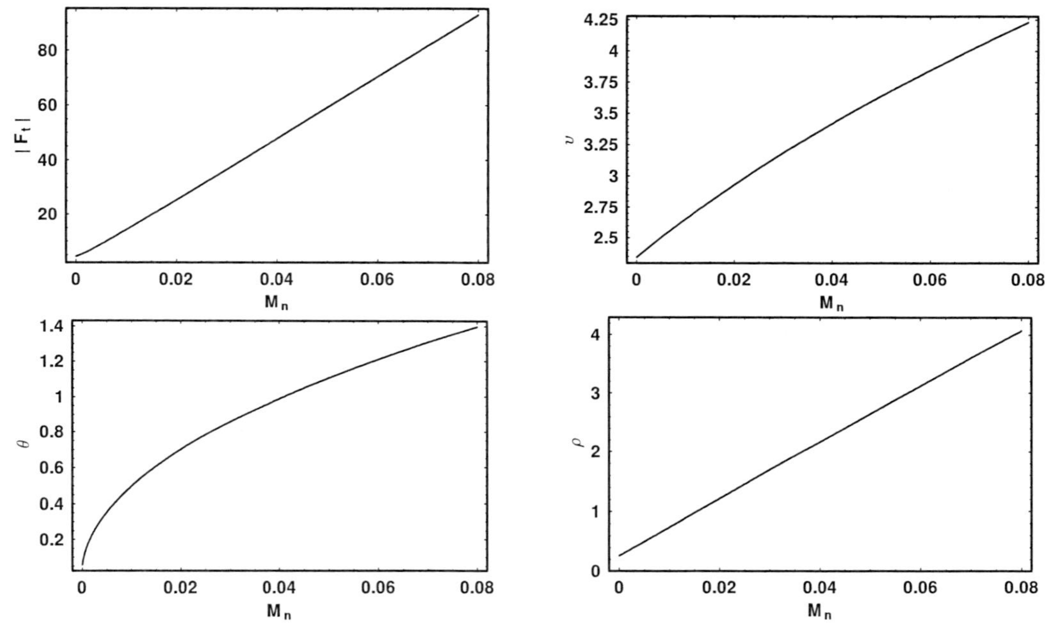

Figure 27. (a-d): Evolution of different dynamical quantities of the system, as a function of the mass of the nucleus M_n. Details are given in the text.

a considerable number of resonance radii for both the direct and retrograde orbits in both cases. Details regarding the resonances and the resonance radii can be obtained form Table 2. In other words, all the Lindblad resonances given in Table 2, are also responsible for the chaotic motion in the primary galaxy. It is also interesting to note that the above resonances produce large chaotic regions for small values of the distance R, while for larger values of R (see Figure 9a-b) the chaotic regions are extremely small, although the resonance radii are still present. This means that in this case the distance between the two galaxies precedence the Lindblad resonances.

In all cases, the primary galaxy is assumed to have a fixed value of total mass

$$M_P = M_d + M_n = 2.08\,, \tag{25}$$

where $0 \leq M_n \leq 0.08$. For a constant mass and distance R the F_x, F_y and F_z forces, per unit mass, at an arbitrary but fixed point $P_0(x_0, y_0, z_0)$ near the center of the primary galaxy, are linear functions of the mass of the nucleus M_n, of the form

$$\begin{aligned} F_x &= -c_1 - k_1 M_n, \\ F_y &= -c_2 - k_2 M_n, \\ F_z &= -c_3 - k_3 M_n, \end{aligned} \tag{26}$$

where (c_1, c_2, c_3) and (k_1, k_2, k_3) are positive constants, while relation (25) has also been taken into account. Figure 27a shows the total $|F_t|$ force as a function of M_n, at $P_0(x_0, y_0, z_0) = (-0.12, 0.02, 0.01)$, when $R = 1.5$. We observe that the total force increases rapidly as the mass of the nucleus M_n increases its value. The presence of the dense and massive nucleus increases the velocity near the center of the primary galaxy. This can be easily shown, if we consider the limiting curve, that is the curve enclosing all the invariant curves in the $x - p_x$ phase plane. This is obtained if we set $y = 0$ in the Jacobi integral (13). Thus, we have

$$p_y^2 = 2\left[E_J - \Phi_t(x)\right] - p_x^2 \geq 0\,. \tag{27}$$

The curve

$$f(x, p_x) = 2\left[E_J - \Phi_t(x)\right] - p_x^2 = 0\,, \tag{28}$$

is a curve on the $x - p_x$ phase plane, called the limiting curve or the Zero Velocity Curve (Z.V.C). The velocity $\upsilon_x = p_x$ at the limiting curve is the total velocity υ. At a fixed point $x = x_0$ the total velocity is

$$\upsilon = \left[2E_J + const + c_4\,(x_0 - x_1)^2 + \frac{2M_n\,(\lambda_1 - \lambda_2)}{\lambda_1 \lambda_2}\right]^{1/2}, \tag{29}$$

where $c_4 < 1$ is a constant, while

$$\begin{aligned} \lambda_1 &= \sqrt{b^2 + (x_0 - x_1)^2 + (a + h)^2}, \\ \lambda_2 &= \sqrt{(x_0 - x_1)^2 + c_n^2}. \end{aligned} \tag{30}$$

As $\lambda_1 > \lambda_2$ the maximum value of the velocity is met when $x_0 = x_1$ (see also Figure 7a-b). it is evident from equations (29) and (30), that υ increases as M_n increases or c_n decreases. This indicates that high velocities are expected in the central regions of galaxies, with dense and massive nuclei. Figure 27b shows a plot of υ as a function of M_n, when $R = 1.5$ and $E_J = -2.6$. We see that the velocity increases as the mass of the nucleus M_n increases its value. In Figure 27c we observe a plot of the rotational velocity Θ of the primary galaxy, as a function of the mass of the nucleus M_n, when $r = r_0 = 0.01$. The exact function is

$$\Theta(M_n) = r_0 \sqrt{\frac{M_n}{\left(r_0^2 + c_n^2\right)^{3/2}} + \frac{M_d}{\left(b^2 + r_0^2 + (a + h)^2\right)^{3/2}}}\,. \tag{31}$$

From Figure 27c we can derive two basic conclusions: (i) In galaxies with massive and dense central concentrations, the rotational velocity has and retains high values near the galactic center. (ii) The more massive and dense the nucleus, the higher is the value of the rotational velocity Θ. Figure 27d presents the evolution of the mass density ρ of the primary galaxy, as a function of the mass of the nucleus M_n at the point $P_0(x_0, y_0, z_0) = (-0.12, 0.02, 0.01)$. It is evident that as the galaxy evolves and a massive nucleus is forming at its center, we have a constant and rapid increase of the mass density. The relationships given in Figure 27a-d are of great importance and will prove very useful in the following section, where we will study the evolution of our galactic system, using a time-dependent model.

6. Evolution of the Orbits in the 3D Time - Dependent Model

Let us now come to follow the evolution of 3D orbits, as mass is transported from the disk of the quiet galaxy to its center. By this procedure a massive and dense nucleus is developed in the central regions of the primary galaxy. The mass transport is linear following the set of equations

$$\begin{aligned} M_{nf}(t) &= M_{ni} + kt, \\ M_{df}(t) &= M_{di} - kt, \end{aligned} \tag{32}$$

where $M_{ni} = 0$ and $M_{di} = 2.08$, are the initial values of the mass of the nucleus and the disk respectively, while k is a positive parameter. We assume that the linear rate described by relations (32) is slow compared to the orbital period of the binary system and therefore it is adiabatic. This is true because the mass transportation period is 100 time units, while the orbital period is about three orders of magnitude smaller. It is also assumed that the transportation stops when the mass of the nucleus in the galactic core takes the value $M_{nf} = 0.08$. It is well known that the shape of the 3D orbits, sometimes is inconclusive or misleading. In order to overcome this drawback, we have decided to use a highly accurate method such as the Lyapunov Characteristic Exponent - L.C.E. The main advantage of this method is that it uses certain and objective numerical thresholds, beyond which we can distinguish between ordered and chaotic motion. On the other hand this method is very time consuming, as it needs time intervals of numerical integration of order of 10^5 time units, in order to give reliable and definitive results regarding the nature of a 3D orbit. But this is a "price" we can afford to pay. Additionally, we shall use a fast but qualitative method, that is the profile's pattern of the total velocity, so as to characterize a 3D orbit. In the following, we present the evolution of four different 3D orbits, as the total mass distribution of the dynamical system changes with time, following the set of equations (32). For all orbits shown in Figures 28–31, the initial value of the Hamiltonian (10), at $t = 0$, is $E_{Ji} = -2.6$, while $k = 0.0008$.

In Figure 28a we can see the evolution of the Lyapunov Characteristic Exponent for a 3D orbit and for a time period of 10^5 time units, as the galaxy evolves following the set of equations (32). The initial conditions are: $x_0 = 0.12$, $y_0 = 0$, $z_0 = 0.02$, $p_{x0} = 0$, $p_{z0} = 0$, while the value of p_{y0} is found from the Hamiltonian (10) in all cases. The values of all the other parameters are as in Figure 7b. When $t = 100$ time units, the mass of the developed nucleus in the core of the galaxy is $M_{nf} = 0.08$ and the evolution stops. The value of the Hamiltonian is now $E_{Jf} = -2.60012$. The profile of the L.C.E clearly indicates that this

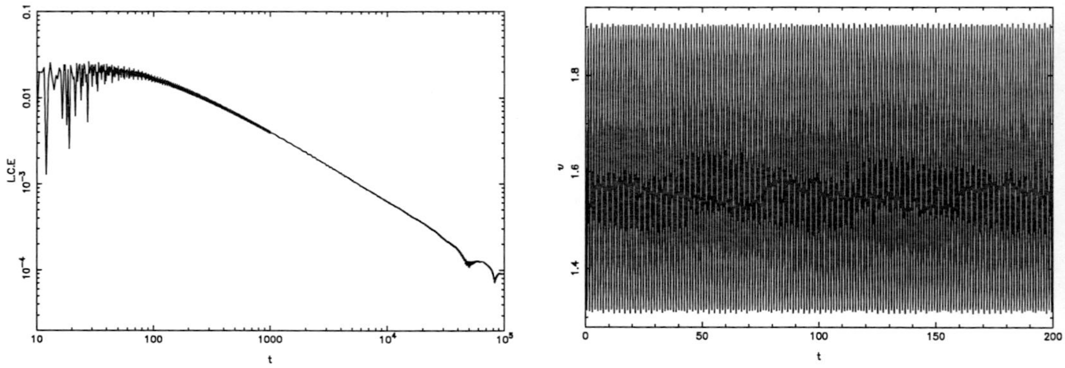

Figure 28. (a-b): (a-left): Evolution of the L.C.E of the 3D orbit in the time-dependent model, following relations (32) and (b-right): The corresponding velocity profile. The orbits starts as a regular and remains regular during the mass transportation.

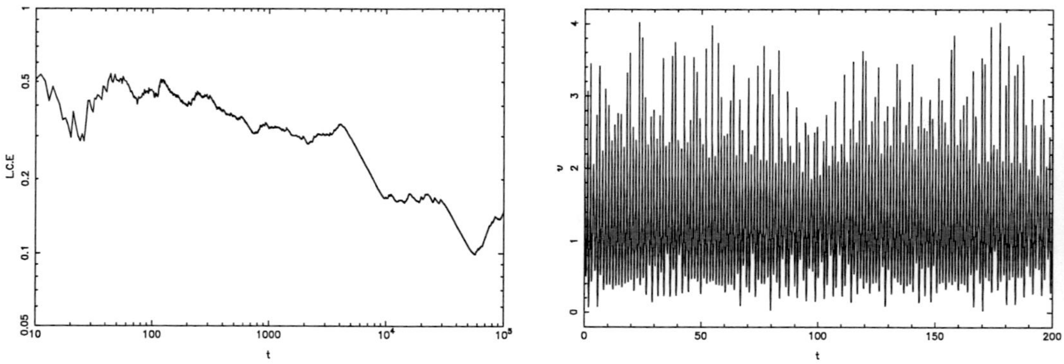

Figure 29. (a-b): (a-left): Evolution of the L.C.E of the 3D orbit in the time-dependent model, following relations (32) and (b-right): The corresponding velocity profile. The orbits starts as a chaotic and remains chaotic during the mass transportation.

orbit starts as a regular and remains regular, during the galactic evolution. Figure 28b shows the velocity profile for the same 3D orbit, for a time period of 200 time units. The pattern is almost quasi-periodic and therefore the orbit remains regular during the galactic evolution.

Things are quite different in Figure 29a. This orbit has initial conditions: $x_0 = -0.44$, $y_0 = 0$, $z_0 = 0.01$, $p_{x0} = 0$, $p_{z0} = 0$. The values of all the other parameters are as in Figure 7b. When $t = 100$ time units, the mass of the developed nucleus in the core of the primary galaxy is $M_{nf} = 0.08$ and the mass transportation stops. The value of the Hamiltonian is now $E_{Jf} = -2.60057$. In this case, the profile of the L.C.E clearly indicates that this orbit starts as a chaotic 3D orbit and remains chaotic, during the galactic evolution. Figure 29b shows the velocity profile for the same 3D orbit, for a time period of 200 time units. The pattern is highly asymmetric, with a lot of abrupt peaks, large deviations between the maxima and also significant deviations between the minima. Therefore, we conclude that this orbit remains chaotic during the mass transportation.

In Figure 30a we observe the evolution of the L.C.E of a 3D orbit with initial conditions: $x_0 = -0.196$, $y_0 = 0$, $z_0 = 0.01$, $p_{x0} = 0$, $p_{z0} = 0$. The values of all the other parameters

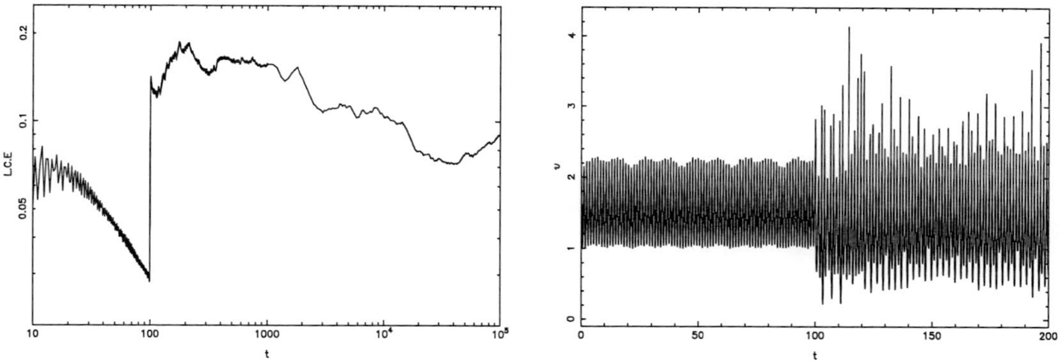

Figure 30. (a-b): (a-left): Evolution of the L.C.E of the 3D orbit in the time-dependent model, following relations (32) and (b-right): The corresponding velocity profile. The orbits starts as a regular but after 100 time units, when the galactic evolution stops it becomes chaotic.

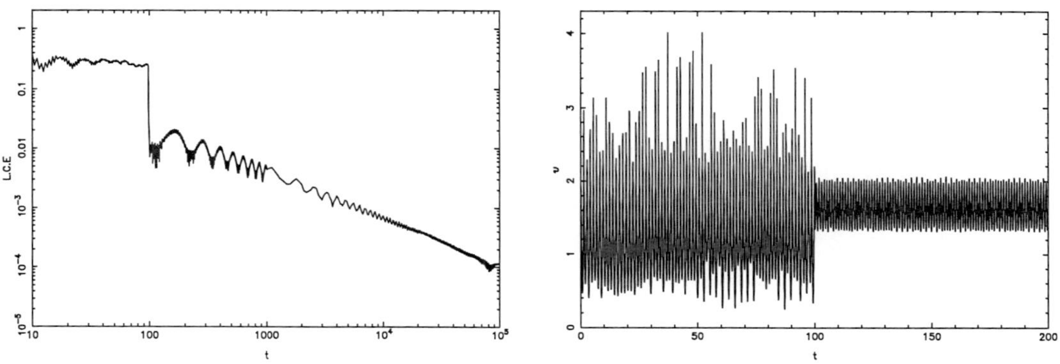

Figure 31. (a-b): (a-left): Evolution of the L.C.E of the 3D orbit in the time-dependent model, following relations (32) and (b-right): The corresponding velocity profile. The orbits starts as a chaotic but after the galactic evolution, it changes its nature to regular.

are as in Figure 7b. After a time interval of 100 time units the mass of the nucleus in the core of the galaxy is $M_{nf} = 0.08$ and the galactic evolution stops. The Hamiltonian settled to the value $E_{Jf} = -2.60034$. The profile of the L.C.E given in Figure 30a shows that the orbit stars as a regular 3D orbit but after the galactic evolution it becomes a chaotic one. It is evident that, if mass transport was not present, the orbit would have remain regular. The presence of the dense and massive nucleus in the core of the galaxy, has changed the character of the 3D orbit, from regular to chaotic. The same result can be obtained by looking the profile of the velocity of the 3D, shown in Figure 30b. For the first 100 time units the pattern is quasi-periodic indicating regular motion, but after that time interval, it becomes highly asymmetric leading to the conclusion that the orbit has become chaotic.

Figure 31a depicts the evolution of the L.C.E of a 3D orbit with initial conditions: $x_0 = 0.74$, $y_0 = 0$, $z_0 = 0.01$, $p_{x0} = 0$, $p_{z0} = 0$. The values of all the other parameters are as in Figure 7b. After a time interval of 100 time units the mass of the nucleus in the core of the galaxy is $M_{nf} = 0.08$ and the mass transportation stops. The Hamiltonian settled to the value $E_{Jf} = -2.60021$. The profile of the L.C.E given in Figure 31a indicates that this orbit stars as a chaotic 3D orbit but after the galactic evolution it becomes an ordered one. Therefore, it is evident that, if mass transportation was not present, the orbit this time would have remain chaotic. In this case, the presence of the nucleus in the core of the galaxy, has changed the nature of the 3D orbit, from chaotic to regular. Similar result can be obtained from the profile of the velocity of the 3D, shown in Figure 31b. For the first 100 time units the pattern is highly asymmetric indicating chaotic motion, but for the rest 100 time units, it becomes almost quasi-periodic leading to the conclusion that the orbit has now become regular.

The above analysis reveals that, as the galaxy tends to create a massive and dense nucleus, the character of most 3D regular orbits, starting near the center of the galaxy, change their nature from regular to chaotic. At the same time, the velocity shows significant changes and increases, as the mass of the nucleus increases. In other words, as the quiet galaxy shows this active face, the nature of motion presents major alterations. Using the above procedure, we have tested a large number of 3D orbits (approximately 1000) in the time-dependent model, describing the formation of massive and dense nucleus in the core of the the primary galaxy, when mass is transported from the disk. Our numerical results indicate that the character of the 3D orbits can change either from regular to chaotic and vise versa or not change at all, as the mass is transported and the massive and dense nucleus is developed in the central region of the disk galaxy. In particular, from the sample of the 1000 tested orbits in the case of the time-dependent model, we conclude that: 61% of the orbits altered their nature from regular to chaotic, 22% remain chaotic, 15% remain regular and only 2% changed their character from chaotic to regular. Thus, it is evident that the formation of a massive nucleus, leads the majority of the orbits of the system to become chaotic. Whether the nature of a 3D orbit will change or not during the galactic evolution described by the set of equations (32), strongly depends on the initial conditions (x_0, p_{x0}, z_0) of each orbit. Moreover, as the change of the value of the Hamiltonian (10) is negligible $\left(\Delta E_J = |E_{Jf} - E_{Ji}| \simeq 10^{-4}\right)$, we can say that the phase space is transformed to itself during the galactic evolution and therefore the orbits can be considered iso-energetic. In order to make this statement more clear would present an example. If we suppose that the evolving time-dependent model was describing the 2D system, then as the change of the Hamiltonian is negligible, we could say that in the phase plane of the quiet galaxy shown in Figure 7b chaotic regions would appear in the central area and it would be transformed to the phase plane shown in Figure 7a, as the nucleus is formed in the core of the disk galaxy.

7. Conclusion

In the present chapter, we have constructed a 3D gravitational model, in order to study the character of motion in a binary stellar system. In particular, we present a dynamical model which is composed of a primary galaxy and a small satellite companion. The motion of a test particle (star), in the gravitational field of this galactic system is studied and various techniques are used in order to identify regular and chaotic motion. The galaxies are assumed to be coplanar and orbiting each other in the same plane on a circular orbit. Obviously, this is a contrived model, but contrived models often can provide insight into more realistic situations.

Two different cases were investigated: the time-independent model and the evolving model, that is the case when mass is transported adiabatically from the disk of the galaxy to its center, forming a massive and dense nucleus. Our numerical calculation indicate that there are several factors responsible for the observed chaotic motion, in the time-independent model: (i) the galactic interaction, (ii) the galactic activity, that is the presence of the nucleus and (iii) the Lindblad resonances. Furthermore, the presence of the nucleus increases the velocities near the central region of the primary galaxy. The value of the velocity depends on the mass of the dense nucleus and also on the value of its scale length. Regular motion corresponds to low central velocities, while chaotic motion is characterized by high velocities. All the above, strongly indicate that in the centers of active galaxies chaotic motion in high velocities is expected. On the other hand, it was found that the two interacting galaxies, for large values of the distance between their centers R, do not present chaotic motion, when the massive nucleus is not present in the core of the primary one.

We have started our investigated from the Hamiltonian system of two degrees of freedom (2D). Our numerical simulation indicate that in this case, a large part of chaotic of the phase plane is covered by chaotic orbits, while the regular regions are confined mainly, near the stable retrograde periodic point. The chaotic area, is larger in the case when the primary galaxy possesses a dense and massive nucleus, or when the distance R between the the two galaxies is small. It was also found, that the velocity near the central region of the galaxy increases significantly, when the dense nucleus is present. For lower values of the total energy and small value of the distance R, no chaos is observed when the nucleus is absent, while a small chaotic region appears in the presence of a massive and dense nucleus. This means, that low energy stars are in chaotic orbits near the centers of active galaxies with a suitable satellite companion in a circular orbit, while in quiet galaxies they are not. Furthermore, the velocity near the center raises to high values, even for smaller values of the energy. All the above, strongly indicate that in the centers of active galaxies, chaotic motion and high velocities are expected. This fact combined with outcomes from previous works, shows that the majority of orbits in galaxies with dense and massive nuclei is chaotic (Caranicolas & Innanen, 1991; Caranicoals & Papadopoulos, 2003). This seems reasonable because theoretical results show that the nuclear force near the center increases linearly as M_n increases. Furthermore, the velocity near the galactic center strongly depends on the scale size of the nucleus c_n. The smaller that c_n is, the higher is the velocity near the central region of an active galaxy.

In order to explore and understand the nature of orbits in the 3D dynamical system, we have used our knowledge obtained from the study of the 2D system. Of particular interest was the determination of the regions of initial conditions in the (x, p_x, z), $p_y > 0$, $(y = p_z = 0)$ phase space that produce regular or chaotic 3D orbits. As the value of p_{y0} was found from the Jacobi integral (10), we have used the same value of E_J, as in the 2D system and took initial conditions (x_0, p_{x0}, z_0) such as (x_0, p_{x0}) lies in the chaotic regions of the 2D system. It was found that the motion is chaotic for all permissible values of z_0. On the other hand, in the case when (x_0, p_{x0}) was inside a regular region around the direct and retrograde periodic points, the corresponding 3D orbits are regular for small values of z_0, while for larger values of z_0 the orbits become chaotic. The particular values of z_0 for which the transition from regularity to chaos in 3D orbits is occurred were different for each regular region of the 2D system. Of particular interest are the results given in Table 1, where we define the average minimum value of z_0 near the direct and retrograde periodic points of the primary galaxy, where the transition from regularity to chaos occurs.

In this research, we have deployed different kinds of theoretical techniques in an attempt to explain and study the dynamical structure of the 2D or 3D Hamiltonian system. An important role is played by the L_z component of the test particle's angular momentum. It was found that the values of $\langle L_z \rangle$ for regular 3D orbits are larger, than those corresponding to chaotic 3D orbits. Thus, the L_z component of the angular momentum is a significant dynamical parameter connected with the regular or chaotic character of orbits in both 2D and 3D dynamical systems. In order to estimate the degree of chaos in the 2D as well as in the 3D dynamical system, we have computed the average value of the Lyapunov Characteristic Exponent – L.C.E, for a large number of orbits with different initial conditions in the chaotic regions in each case, for a time period of 10^5 time units. Note that all the calculated L.C.Es were different on the fifth decimal point in the same chaotic region. Therefore, our numerical results indicate that the degree of chaos in the 3D binary stellar system of interacting galaxies, is smaller than in similar 2D systems.

It is of great interest to follow the evolution of orbits as the massive and dense nucleus is formed in the central region of the primary disk galaxies. In this procedure, mass is transported from the disk to the nucleus and therefore the quiet galaxy become gradually active, following equations (32). We observe that the final character of the 3D orbits strongly depends on the particular initial conditions (x_0, p_{x0}, z_0). Thus, regular orbits can turn to chaotic and vice versa, or maintain their character (ordered or chaotic) during the nuclear formation and since the massive nucleus has been formed. It was observed, that a number of regular quasi-periodic orbits, starting near the central region of the primary galaxy, become chaotic. This can been seen in a total velocity vs. time plot, where the asymmetric profile of the total velocity is in agreement with observational data (see Grosbøl, 2002), where an increase of the stellar velocity is expected, in regions with significant chaoticity. Moreover, observations show that an asymmetric velocity profile indicates chaotic motion. An interesting question is whether interactions is essential in order to trigger the mass transportation and therefore the galactic activity. This question remains unsolved as astronomers have found binary systems with disk galaxies (see Letawe et al., 2006) showing no signs of interaction at all but harboring active nuclei.

Our theoretical dynamical analysis of the binary stellar system, can now be compared with observational data derived from M51-type binary galactic systems. These systems consist of a large, primary, disk galaxy and a smaller satellite companion galaxy moving in a circular orbit around their common mass center (see Klimanov & Reshetnikov, 2001). We apply our results, to the binary stellar system composed of the primary galaxy NGC 5829 and its small satellite IC 4526. Figure 32 depicts a real image of this binary stellar system. The primary galaxy has an absolute magnitude $M_V = -20.66$. In order to estimate the disk's radius, we shall use the empirical relation

$$M_V = -6logA + 7.14\,, \tag{33}$$

which connects the the absolute magnitude to the projected major axis of the primary galaxy A in pc (see Bowers & Deeming, 1984). From relation (33) we obtain the value $A = 43$ kpc. If we adopt this values of the parameters given in Figure 7a-b, then one should expects a large number of the orbits to be chaotic in this binary system of interacting galaxies. We

Figure 32. A real image of the binary stellar system, composed of the primary galaxy NGC 5829 and its small satellite companion IC 4526.

can also observe that the orbit of the satellite is close to the boundary of the galactic disk. This result is in agreement with the observational data as well.

When studying Galactic Dynamics, it is invariably found that the stellar rotation velocity remains constant, or flat, with increasing distance away from the galactic center (see Catinella et al. 2006). This result, is highly counterintuitive since, based on Newton's Law of gravity, the rotational velocity would steadily decrease for stars further away from the galactic center. By this particular argument, the flat rotational curves, seem to imply that each galaxy must be surrounded by significant amounts of dark matter. It has been postulated and generally accepted, that the dark matter would have to be located in a massive, halo enshrouding each galaxy (see Caranicolas & Zotos, 2009; Caranicolas & Zotos, 2010; Caranicolas & Zotos, 2011). The first real surprise in the study of dark matter lay in the outermost parts of galaxies, known as galaxy halos. Here, there is negligible luminosity, yet there are occasional orbiting gas clouds, which allow one to measure rotation velocities and distances. The rotation velocity is found, not to decrease with increasing distance from the galactic center, implying that the mass distribution of the galaxy, cannot be considered, like the light distribution. The mass must continue to increase: since the rotation velocity satisfies

$$\Theta^2 = \frac{GM}{r}, \tag{34}$$

where M is the mass within radius r, we infer that M increases proportionally to r. This rise appears to stop at about $50kpc$, where halos appear to be truncated. We infer that the mass-luminosity ratio of the galaxy (M/L), including its disk-halo, is about 5 times larger than estimated for the luminous inner region, or equal to about 50. The rotation curve shown in Figure 3, corresponding to our disk galaxy seems to maintain high values of the angular velocity Θ for large distances from the galactic center. Thus, we may say that our dynamical model can interpret the presence of dark matter.

Forty years ago, galactic activity and interactions between galaxies were viewed as unusual and rare. Nowadays, they seem to be segments in the life of many galaxies. From the astrophysical point of view, in the present work, we have tried to connect galactic activity and galactic interactions with the nature of orbits (regular or chaotic) and also with the behavior of the velocities of stars in the primary galaxy. Here, we must point out that a binary stellar system of interacting galaxies is very complex and therefore, we need to assume some necessary simplifications and assumptions, in order to be able to study the orbital behavior of such a complex system. We consider the outcomes of the present research, to be an initial effort, in order to explore the dynamical structure of the 3D binary stellar system, in more detail. As results are positive, further investigation will be initiated to study all the available phase space, including orbital eccentricity of the small companion galaxy (the galaxy with the smallest value of the total mass) and its inclinations to the primary galaxy.

References

[1] Abadi, M. G., Navarro, J. F., Steinmetz, M. 2006, *MNRAS* **365**, 747.

[2] Baade, W. 1944, *ApJ* **100**, 137.

[3] Barker, M. K., Ferguson, A. M. N., Irwin, M., Arimoto, N., Jablonka, P., 2009, *AJ* **138**, 1469.

[4] Bekki, K., Chiba, M. 2001, *ApJ* **558**, 666.

[5] Bekki, K., Chiba, M. 2005, *MNRAS* **356**, 680.

[6] Bell, E. F., Zucker, D. B., Belokurov, V., Sharma, S., Johnston, K. V., Bullock, J. S., Hogg, D. W., Jahnke, K., et al. 2008, *ApJ* **680**, 295.

[7] Belokurov, V., Evans, N. W., Bell, E. F., Irwin, M. J., Hewett, P. C., Koposov, S., Rockosi, C. M., Gilmore, G., et al. 2007a, *ApJ* **657**, L89.

[8] Belokurov, V., Evans, N. W., Irwin, M. J., Lynden-Bell, D., Yanny, B., Vidrih, S., Gilmore, G., Seabroke, G., et al. 2007b, *ApJ* **658**, 337.

[9] Belokurov, V., Zucker, D. B., Evans, N. W., Gilmore, G., Vidrih, S., Bramich, D. M., Newberg, H. J., Wyse, R. F. G., et al. 2006, *ApJ* **642**, L137.

[10] Binney, J., Tremaine, Sc. 2008, *Galactic Dynamics*, Princeton Series in Astrophysics, 2nd edition.

[11] Bower, R. G., Benson, A. J., Malbon, R., Helly, J. C., Frenk, C. S., Baugh, C. M., Cole, S., Lacey, C. G. 2006, *MNRAS* **370**, 645.

[12] Bowers, R. L. Deeming, T. 1984, *Astrophysics II*, Jones and Bartlett Publishers, Inc., Boston.

[13] Brook, C. B., Kawata, D., Gibson, B. K., Flynn, C. 2004, *MNRAS* **349**, 52.

[14] Byrd, G. 1978, *ApJ* **226**, 70.

[15] Cantinella, B., Giovanelli, R., Haynes, M. P. 2006, *AJ* **640**, 751.

[16] Carollo, D., Beers, T. C., Chiba, M., Norris, J. E., Freeman, K. C., Lee, Y. S., Ivezic, Z., Rockosi, C. M., et al. 2009, ArXiv e-prints astroph/0909.3019.

[17] Carollo, D., Beers, T. C., Lee, Y. S., Chiba, M., Norris, J. E., Wilhelm, R., Sivarani, T., Marsteller, B., et al. 2007, *Nature* **450**, 1020.

[18] Caranicolas, N. D., Innanen, K. A. 1991, *AJ* **102**, 1343.

[19] Caranicolas, N. D., Papadopoulos, N. J. 2003, *A&A* **399**, 957.

[20] Caranicolas, N. D., Papadopoulos, N. J. 2009, *New Astronomy* **14**, 207.

[21] Caranicolas, N. D., Vozikis, Ch. L. 1999, *A&A* **349**, 70.

[22] Caranicolas, N. D., Zotos, E. E. 2009, *Mechanics Research Communications* **36**, 875.

[23] Caranicolas, N. D., Zotos, E. E. 2010, *New Astronomy* **15**, 427.

[24] Caranicolas, N. D., Zotos, E. E. 2011, *Research in Astronomy and Astrophysics* **11**(7), 811.

[25] Cincotta, P. M., Giordano, C. M., Perez, M. J. 2006, *A&A* **455**, 499.

[26] Connors, T. W., Kawata, D., Maddison, S. T., Gibson, B. K. 2004, *PASA* **21**, 222.

[27] Contopoulos, G., Magnenat, P., Martinet, L. 1982, *Physica D* **6**, 123.

[28] Contopoulos, G., Grøsbol, P. 1986, *A&A* **155**, 11.

[29] Contopoulos, G., et al. 1995, *A&A* **304**, 347.

[30] Contopoulos, G., et al. 1997, *CeMDA* **67**, 293.

[31] Contopoulos, G., Voglis, N. 1996, *CeMDA* **64**, 1.

[32] Croton, D. J., Springel, V., White, S. D. M., De Lucia, G., Frenk, C. S., Gao, L., Jenkins, A., Kauffmann, G., et al. 2006, *MNRAS* **365**, 11.

[33] de Jong, R. S., Radburn-Smith, D. J., Sick, J. N. 2008, in Astronomical Society of the Pacific Conference Series, Vol. 396, *Formation and Evolution of Galaxy Disks*, Funes, J. G., Corsini, E. M., eds. p. 187.

[34] De Lucia, G., Springel, V., White, S. D. M., Croton, D., Kauffmann, G., 2006, *MNRAS* **366**, 499.

[35] Eggen, O. J., Lynden-Bell, D., Sandage, A. R. 1962, *ApJ* **136**, 748.

[36] Faúndez-Abans, M., Reshetnikov, V. P., de Oliveira-Abans, M., Fernandes, I. F. 2009, *Astronomy Letters* **35**, 25.

[37] Ferguson, A.M. N., Irwin, M. J., Ibata, R. A., Lewis, G. F., Tanvir, N. R., 2002, *AJ* **124**, 1452.

[38] Freeman, K., Bland-Hawthorn, J. 2002, *ARA&A* **40**, 487.

[39] Froeschlé, C. 1970, *A&A* **4**, 115.

[40] Froeschlé, C. 1972, *A&A* **16**, 172.

[41] Froeschlé, C., et al. 1993, *CeMDA* **56**, 307.

[42] Gardiner, L. T., Sawa, T., Fujimoto, M. 1994, *MNRAS* **278**, 191.

[43] Grøsbol, P. 2002, *SSRv* **102**, 73.

[44] Hasan, H., Norman, C. A. 1990, *ApJ* **361**, 69.

[45] Hasan, H., et al. 1993, *ApJ* **409**, 91.

[46] Hatton, S., Devriendt, J. E. G., Ninin, S., Bouchet, F. R., Guiderdoni, B., Vibert, D., 2003, *MNRAS* **343**, 75.

[47] Helmi, A. 2008, *A&A Reviews* **15**, 145.

[48] Helmi, A., White, S. D. M. 1999, *MNRAS* **307**, 495.

[49] Henrard, J., Caranicolas, N. D. 1990, *CeMDA* **47**, 99.

[50] Ibata, R., Irwin, M., Lewis, G. F., Stolte, A. 2001, *ApJ* **547**, L133.

[51] Ibata, R., Martin, N. F., Irwin, M., Chapman, S., Ferguson, A. M. N., Lewis, G. F., McConnachie, A. W. 2007, *ApJ* **671**, 1591.

[52] Ibata, R., Mouhcine, M., Rejkuba, M. 2009, *MNRAS* **395**, 126.

[53] Ibata, R. A., Gilmore, G., Irwin, M. J. 1994, *Nature* **370**, 194.

[54] Johnston, K. V., Hernquist, L., Bolte, M. 1996, *ApJ* **465**, 278.

[55] Jurić, M., Ivezić, v. Z., Brooks, A., Lupton, R. H., Schlegel, D., Finkbeiner, D., Padmanabhan, N., Bond, N., et al. 2008, *ApJ* **673**, 864.

[56] Kalirai, J. S., Gilbert, K. M., Guhathakurta, P., Majewski, S. R., Ostheimer, J. C., Rich, R. M., Cooper, M. C., Reitzel, D. B., et al. 2006, *ApJ* **648**, 389.

[57] Kang, X., Jing, Y. P., Mo, H. J., Börner, G., 2005, *ApJ* **631**, 21.

[58] Karanis, G. I., Caranicolas, N. D. 2001, *A&A* **367**, 443.

[59] Karanis, G. I., Caranicolas, N. D. 2002, *Astron. Nachr.* **323**, 3.

[60] Kauffmann, G., Colberg, J. M., Diaferio, A., White, S. D. M. 1999, *MNRAS* **303**, 188.

[61] Kauffmann, G., Nusser, A., Steinmetz, M. 1997, *MNRAS* **286**, 795.

[62] Klimanov, S. A., Reshentikov, V. P. 2001, *A&A* **378**, 428.

[63] Laird, J. B., Carney, B. W., Rupen, M. P., Latham, D. W. 1988, *AJ* **96**, 1908.

[64] Letawe, G., et al. 2006, *New Astronomy Reviews* **50**, 779.

[65] Lichtenberg, A. J., Lieberman, M. A. 1992, *Regular and Chaotic Dynamics*, Springer, 2nd edition.

[66] Lin, D. N. C., Jones, B. F., Klemola, A. R. 1995, *ApJ* **439**, 652.

[67] Magnenat, P. 1982, *Celestial Mechanics* **28**, 319.

[68] Malin, D., Hadley, B., 1999, in Astronomical Society of the Pacific Conference Series, Vol. 182, *Galaxy Dynamics* - A Rutgers Symposium, D. R. Merritt, M. Valluri, J. A. S., ed. p. 445.

[69] Martinet, L., Magnenat, P. 1981, *A&A* **96**, 68.

[70] Martínez-Delgado, D., Peñarrubia, J., Gabany, R. J., Trujillo, I., Majewski, S. R., Pohlen, M. 2008, *ApJ* **689**, 184.

[71] Martínez-Delgado, D., Pohlen, M., Gabany, R. J., Majewski, S. R., Peñarrubia, J., Palma, C. 2009, *ApJ* **692**, 955.

[72] McConnachie, A. W., Chapman, S. C., Ibata, R. A., Ferguson, A. M. N., Irwin, M. J., Lewis, G. F., Tanvir, N. R., Martin, N. 2006, *ApJ* **647**, L25.

[73] McConnachie, A. W., Irwin, M. J., Ibata, R. A., Dubinski, J., Widrow, L. M., Martin, N. F., Côté, P., Dotter, A. L., et al. 2009, *Nature* **461**, 66.

[74] Merritt, D. 1996, *Science* **271**, 337.

[75] Miyamoto, W., Nagai, R. 1975, *PASJ* **27**, 533.

[76] Morrison, H. L., Helmi, A., Sun, J., Liu, P., Gu, R., Norris, J. E., Harding, P., Kinman, T. D., et al. 2009, *ApJ* **694**, 130.

[77] Morrison, H. L., Mateo, M., Olszewski, E. W., Harding, P., Dohm-Palmer, R. C., Freeman, K. C., Norris, J. E., Morita, M. 2000, *AJ* **119**, 2254.

[78] Newberg, H. J., Yanny, B., Rockosi, C., Grebel, E. K., Rix, H. W., Brinkmann, J., Csabai, I., Hennessy, G., et al. 2002, *ApJ* **569**, 245.

[79] Oort, J. H. 1926, *Publications of the Kapteyn Astronomical Laboratory Groningen* **40**, 1.

[80] Papadopoulos, N. J., Caranicolas, N. D. 2006, *New Astronomy* **12**, 11.

[81] Patsis, P. A., et al. 1997, *A&A* **326**, 493.

[82] Pfenniger, D. 1984, *A&A* **134**, 373.

[83] Patsis, P. A., Zachilas, L. 1994, *Int. J. Bif. Chaos* **4**, 1399.

[84] Putman, M. E., Staveley-Smith, L., Freeman, K. C., et al. 2003, *ApJ* **586**, 170.

[85] Quinn, P. J. 1984, *ApJ* **279**, 596.

[86] Revaz, Y., Pfenniger, D. 2001, *A&A* **372**, 784.

[87] Sackett, P. D., Morrisoni, H. L., Harding, P., Boroson, T. A. 1994, *Nature* **370**, 441.

[88] Saito, N., Ichimura, A., 1979, In Casati, G. & Ford, L. (eds) *Stochastic Behaviour in Classical and Quantum Hamiltonian Systems*, Springer, Berlin, p.137.

[89] Schlaufman, K. C., Rockosi, C. M., Beers, T. C., Bizyaev, D., Brewington, H., Lee, Y. S., Malanushenko, V., Malanushenko, E., et al. 2009, ArXiv e-prints astro-ph/0908.2627.

[90] Schweizer, F. 1980, *ApJ* **237**, 303.

[91] Searle, L., Zinn, R. 1978, *ApJ* **225**, 357.

[92] Sellwood, J. A., Moore, E. M. 1999, *ApJ* **510**, 125.

[93] Shang, Z., Zheng, Z., Brinks, E., Chen, J., Burstein, D., Su, H., Byun, Y. I., Deng, L., et al. 1998, *ApJ* **504**, L23.

[94] Skokos, Ch., Contopoulos, G., Polymilis, C. 1997, *CeMDA* **65**, 223.

[95] Skokos, Ch., Contopoulos, G., Polymilis, C. 1999, in *Hamiltonian Systems with three or more degrees of freedom*, ed. by Simó C., Plenum Press, p. 583-587.

[96] Springel, V., White, S. D. M., Jenkins, A., Frenk, C. S., Yoshida, N., Gao, L., Navarro, J., Thacker, R., et al. 2005, *Nature* **435**, 629.

[97] Springel, V., White, S. D. M., Tormen, G., Kauffmann, G. 2001, *MNRAS* **328**, 726.

[98] Starkenburg, E., Helmi, A., Morrison, H. L., Harding, P., van Woerden, H., Mateo, M., Olszewski, E.W., Sivarani, T., et al. 2009, *ApJ* **698**, 567.

[99] Van der Marel, R. P. 2001, *AJ* **122**, 1827.

[100] Vivas, A. K., Zinn, R. 2006, *AJ* **132**, 714.

[101] Voglis, N., Contopoulos, G. 1994, *J. Phys. A.* **27**, 4899.

[102] Vozikis, Ch. L., Caranicolas, N. D. 1992, *JA&A* **13**, 319.

[103] Vozikis, Ch. L., Caranicolas, N. D. 1994, *A&A* **288**, 448.

[104] Watkins, L. L., Evans, N. W., Belokurov, V., Smith, M. C., Hewett, P. C., Bramich, D. M., Gilmore, G. F., Irwin, M. J., et al. 2009, *MNRAS* **398**, 1757.

[105] Weinberg, M. D. 1995, *ApJ* **455**, L31.

[106] Weinberg, M. D. 1998, *MNRAS* **299**, 499.

[107] Wisdom, J. 1987, *AJ* **94**, 1350.

[108] Yanny, B., Newberg, H. J., Kent, S., Laurent-Muehleisen, S. A., Pier, J. R., Richards, G. T., Stoughton, C., Anderson, Jr. J. E., et al. 2000, *ApJ* **540**, 825.

[109] Yanny, B., Newberg, H. J., Grebel, E. K., Kent, S., Odenkirchen, M., Rockosi, C. M., Schlegel, D., Subbarao, M., et al. 2003, *ApJ* **588**, 824.

[110] Yanny, B., Rockosi, C., Newberg, H. J., Knapp, G. R., Adelman-McCarthy, J. K., Alcorn, B., Allam, S., Allende, Prieto, C., et al. 2009, *AJ* **137**, 4377.

[111] Zibetti, S., Ferguson, A. M. N. 2004, *MNRAS* **352**, 64.

[112] Zolotov, A., Willman, B., Brooks, A. M., Governato, F., Brook, C. B., Hogg, D. W., Quinn, T., Stinson, G. 2009, *ApJ* **702**, 1058.

[113] Zotos, E. E. 2011a, *New Astronomy* **16**, 391.

[114] Zotos, E. E. 2011b, *Baltic Astronomy* **20**, 77.

[115] Zotos, E. E. 2011c, *Chaos, Solitons & Fractals* **44**, 501.

In: Galaxies: Evolution, Morphology and Dynamics ISBN: 978-1-62081-185-6
Editors: A. Guerrera Sanchez et al., pp. 111-133

Chapter 4

AGE VS. OXYGEN ABUNDANCE AND OXYGEN ABUNDANCE DISTRIBUTION IN NEARBY STARS: THE SCATTER DUE TO STOCHASTIC STAR FORMATION

Roberto Caimmi*
Dipartimento di Astronomia
Università di Padova, Padua, Italy

Abstract

A large dispersion observed in the age-metallicity relation inferred from different populations within the solar neighbourhood, is interpreted as largely due to intrinsic scatter and compared to predictions from inhomogeneous simple models of chemical evolution with stochastic star formation (Caimmi 2000). More specifically, the interstellar medium is idealized as substructured into a discrete number of n identical active or quiescent regions where star formation can or cannot randomly occur, respectively. The evolution is idealized as a series of identical time steps where stellar nucleosynthesis takes place within active regions, and subdivided into an assembling stage where the volume undergoes substantial changes, and a subsequent relaxed stage where the volume remains more or less unaltered. At the end of any step, a mass fraction (including newly synthesised elements) from short-lived stars is returned to the interstellar medium, instantaneously mixed with remaining gas from active and quiescent regions, and a new set of identical n regions is made for the next step. The theoretical age-oxygen abundance relation (TAOR) and the theoretical differential oxygen abundance distribution (TDOD) are computed for three different runs where the state of a region (active or quiescent) is decided by different sequences of random numbers. For each Galactic environment, different cases involving low ($n = 10$) and high ($n = 100$) number of regions are considered together with the case of expected evolution ($n \to +\infty$) exploited in an earlier attempt (Caimmi and Milanese 2009). Finally, the TAOR and the TDOD are compared with the empirical age-oxygen abundance relation (EAOR) and the empirical differential oxygen abundance distribution (EDOD), respectively, inferred from subsamples made of halo, bulge, thick disk, thin disk stars

*E-mail address: roberto.caimmi@unipd.it

and, in addition, thick + thin disk stars calculated for selected thick to thin disk mass ratios. The main results are (1) the TAOR fits to an acceptable extent (for at least one computer run) to the EAOR for both the halo, the thick disk, and the thick + thin disk, regardless of the number of regions, while the same holds for the thin disk only if the number of regions is low i.e. via stochastic effects, and (2) the TDOD fits to an acceptable extent (for at least one computer run) to the EDOD for all the above mentioned environments, regardless of the number of regions, where the fit is further improved for the thick + thin disk if the number of regions is low i.e. via stochastic effects. The results need to be confirmed using high-population, complete, and unbiased samples. An interpretation of the model within current ΛCDM cosmological scenarios is shortly outlined.

PACS 98.35.Ac; 98.35.Bd; 98.35.Ln

Keywords: Galaxies: evolution - Stars: formation; evolution.

1. Introduction

The empirical age-metallicity relation and the empirical metallicity distribution of long-lived stars of different populations, constrain models for the formation and the evolution of the Galaxy. Simple models make a useful tool in the description of galactic chemical evolution, where the assumption of instantaneous recycling implies the restriction to elements synthesised by a single population (such as oxygen yielded by SNII supernovae), leaving aside elements synthesised by different populations (such as iron yielded by both SNII and SNIa supernovae). Unfortunately, oxygen is more difficult than iron to detect, and only recently direct oxygen abundance determinations have been performed on stars belonging to different populations (e.g., Ramirez et al. 2007, hereafter quoted as Ral07; Melendez et al. 2008) where sample populations amount to a few tens or hundreds at most.

The empirical age-metallicity relation is considered a poor constraint to the Galactic chemical evolution, due to the lack of a clear trend and to the presence of a large scatter (e.g., Edvardsson et al. 1993), which could be intrinsic to the measure procedure and/or the data and/or spurious effects such as sample selection for metallicity (Garnett and Kobulnicky 2000) and biases in isochrone dating methods (Pont and Eyer 2004) or in atmospheric parameters (Haywood 2006). In particular, considerable scatter on the age-metallicity relation results from an assumed 0.1 dex dispersion in [Fe/H] and 50 K in effective temperature (Haywood 2006, Fig. 5 therein). The superposition of systematic biases in [Fe/H] and $T_{\rm eff}$ up to ∓ 0.2 dex and ∓ 200 K, respectively, makes a shift with respect to the theoretical age-metallicity relation (Haywood 2006, Fig. 6 therein).

The scatter intrinsic to the measure procedure, detected on abundances in isotopic ratios of meteoritic presolar dust grains (Nittler 2005) and interstellar medium (Cartlidge et al. 2006) amounts to about 0.04 dex, which is substantially lower with respect to a value of about 0.2 dex due to intrinsic scatter or biases in atmospheric parameters. Then the remaining alternative is that the scatter in metal abundance is intrinsic to coeval stars, which shall be exploited in the current attempt.

According to current ΛCDM cosmological scenarios, galaxies are built up via hierarchical clustering i.e. less massive systems first form and subsequently merge (or are accreted) to yield more massive systems. Compact dwarf galaxies may safely be conceived as building blocks of larger elliptical and spiral galaxies. Star formation is enhanced by tidal interaction due to merger or accretion events (Rieke et al. 1980; Doane and Mathews 1993; Lopez-Sanchez 2010), implying a more rapid metal enrichment in the interstellar medium with respect to isolated systems. Accordingly, coeval stars within galaxies which underwent accretion and/or merger events are expected to exhibit a large scatter in metal abundance. On the other hand, the gas left in the accreted and in the accreting system is expected to be homogeneously mixed. From an astrophysical standpoint, a similar process may act on subunits belonging to massive galaxies: cloud collisions or close encounters in ellipticals (e.g., Brosce 1970; Pilyugin 1996) and density waves in spirals (e.g., Lin and Shu 1964; Lin et al. 1969).

In this view, star formation may be conceived as inhomogeneous in the sense that it occurs within a fixed fraction of the total gas mass, where the remaining gas is returned to the intercloud medium after an assigned step (Malinie et al. 1993). In addition, star formation may be conceived as stochastic, in the sense that it cannot be established where and when star formation shall take place (Wilmes and Köppen 1995). In earlier attempts (Caimmi 2000 + erratum 2001a, hereafter quoted together as C00; Caimmi 2001b, hereafter quoted as C01; Caimmi 2007, hereafter quoted as C07), the theoretical age-oxygen abundance relation (hereafter quoted as TAOR) and the theoretical differential oxygen abundance distribution (hereafter quoted as TDOD) were fitted to the data for different Galactic environments, by use of empirical [O/H]-[Fe/H] relations. Simple inhomogeneous models considered therein imply star formation (i) within discrete regions, and (ii) via stochastic processes.

The use of simple models allows a deep insight on the influence of input parameters on evolution, and their connection with output parameters. Quoting a classical paper (Pagel 1989):

> "The great virtue of a simple model is that it enables one to form a quick impression of the effects of different physical processes, relevant to the problem under consideration, on the basis of simple analytical formulae without becoming logged down in the wealth of opaque detail that besets numerical calculations and makes it difficult to identify what are the important factors in producing a given result."

In dealing with simple models, it should be looked at their ability to provide a convenient parametrization of the extent to which their assumptions fail, rather than at their ability to describe a dataset.

The empirical age-oxygen abundance relation (hereafter quoted as EAOR) and the differential oxygen abundance distribution (hereafter quoted as EDOD) within long-lived stars of a selected spectral class, can be inferred from recent determinations on a sample of stars belonging to different populations (Ral07). The comparison with related theoretical counterparts was made for the special case of the expected evolution (constant number ratio of active to quiescent regions and infinite total number of regions) in earlier attempts (Caimmi 2008; Caimmi and Milanese 2009, hereafter quoted as CM09). In the current paper, a large dispersion observed in the age-metallicity relation inferred from different populations within the solar neighbourhood, is interpreted as largely due to intrinsic scatter and compared to predictions from inhomogeneous simple models of chemical evolution with stochastic star formation and different number of regions.

Subsamples extracted from a parent sample (Ral07) for different populations (halo, thick disk, thin disk, uncertain) with regard to EAOR and EDOD, are briefly described in Section 2 The model is shortly reviewed in Section 3 The results are presented in Section 4 The discussion is drawn in Section 5 The conclusion is shown in Section 6

2. Data and Inferred Quantities

The EAOR is deduced from a subsample ($N = 222$) of solar neighbourhood disk and halo stars for which ages have been determined (Ral07), hereafter quoted as R07T subsample, from which four subsamples are extracted following the prescriptions of the authors, related to thick disk ($N = 48$), thin disk ($N = 145$), halo ($N = 7$), and uncertain population ($N = 22$) stars, hereafter quoted as R07KT, R07NT, R07HT, and R07UT subsample, respectively.

The EDOD is deduced from a subsample ($N = 522$) of solar neighbourhood disk and halo stars for which oxygen abundances have been determined (Ral07), hereafter quoted as R07O subsample, from which four subsamples are extracted following the prescriptions of the authors, related to thick disk ($N = 133$), thin disk ($N = 310$), halo ($N = 28$), and uncertain population ($N = 51$) stars, hereafter quoted as R07KO, R07NO, R07HO, and R07UO subsample, respectively.

R07K subsamples are biased against low ([Fe/H] $\lesssim -1$) metallicities, and R07H subsamples are incomplete (Ral07).

The EDOD is also deduced from a subsample ($N = 19$) of bulge stars for which oxygen abundances have been determined (Melendez et al. 2008), hereafter quoted as the M08BO subsample.

For further details refer to the parent papers (Ral07; Melendez et al. 2008) and to an earlier attempt (CM09).

The oxygen abundance, $Z_{\rm O}$, shall be normalized to the solar value, as (Pagel, 1989; Malinie et al., 1993; Rocha-Pinto and Maciel, 1996; C00, C01, C07, CM09):

$$\log \phi = \log \frac{Z_{\rm O}}{(Z_{\rm O})_\odot} = \left[\frac{\rm O}{\rm H}\right] ; \tag{1}$$

which holds to a good extent.

The EDOD for stars of a selected spectral class is defined as (Rocha-Pinto and Maciel, 1996; C00, C01, C07, CM09):

$$\psi = \log \frac{\Delta N}{N \Delta \phi} ; \tag{2}$$

with an assumed Poissonian error expressed as (e.g., Ryan and Norris, 1991; C01, C07, CM09):

$$\Delta^{\mp}\psi = \log \left[1 \mp \frac{(\Delta N)^{1/2}}{\Delta N}\right] ; \tag{3}$$

where ΔN is the population of a selected bin, $\phi \mp \Delta\phi/2$.

Table 1. The empirical, differential oxygen abundance distribution (EDOD) in the solar neighbourhood stars of uncertain population, deduced from the R07UO subsample ($N = 51$) in presence of the local thermodynamic equilibrium approximation (LTE).

ϕ	ψ	$\Delta^-\psi$	$\Delta^+\psi$	ΔN
3.183 E−1				
4.008 E−1	−6.707 E−1	+∞	3.010 E−1	1
5.045 E−1				
6.351 E−1	−2.687 E−1	3.010 E−1	1.761 E−1	4
7.996 E−1	−2.718 E−1	2.574 E−1	1.605 E−1	5
1.007 E−0	−2.926 E−2	2.279 E−1	1.487 E−1	6
1.267 E−0	−1.707 E−1	1.651 E−1	1.193 E−1	10
1.595 E−0	−9.465 E−2	1.297 E−1	9.975 E−2	15
2.009 E−0	−4.677 E−1	1.895 E−1	1.315 E−1	8
2.529 E−0	−1.471 E−0	+∞	3.010 E−1	1
3.183 E−0	−1.571 E−0	+∞	3.010 E−1	1

The EDOD related to the R07KO, R07NO, M08BO, R07HO subsamples were listed in an earlier attempt (CM09, Tables 2, 4, 5, 6 therein), where an interested reader is addressed, while the case related to the R07UO subsample is shown in Tables 1 and 2. For R07 subsamples, oxygen abundance has been determined both in presence and in absence of the local thermodynamic equilibrium (LTE) approximation. For further details refer to the parent paper (Ral07).

The EDOD related to the thick + thin disk can be inferred from the data under the assumption of universal stellar initial mass function, for a selected thick to thin disk mass ratio. For further details refer to the parent paper (CM09).

3. The Model

The inhomogeneous model of chemical evolution used in the current attempt has been formulated in an earlier paper (C00) and improved in subsequent investigations (C01; C07), to which an interested reader is addressed for deeper insight. The evolution occurs as a series of identical time steps, and it is conventionally subdivided into a short early stage where assembling and relaxation occur, and a comparable or longer later stage where the system is thought to be relaxed.

At the beginning of each step, a gas subsystem with uniform composition lies within a background of long-lived stars and inert (i.e. prevented from forming stars) gas. The gas subsystem is instantaneously substructured into an assigned number of identical regions, where star formation stochastically takes place in a number of “active” regions while the others remain “quiescent”.

At the end of each step, according to the instantaneous recycling approximation, short-

Table 2. The empirical, differential oxygen abundance distribution (EDOD) in the solar neighbourhood stars of uncertain population, deduced from the R07UO subsample ($N = 51$) in absence of the local thermodynamic equilibrium approximation (NLTE).

ϕ	ψ	$\Delta^-\psi$	$\Delta^+\psi$	ΔN
3.183 E−1	−5.707 E−1	+∞	3.010 E−1	1
4.008 E−1				
5.045 E−1	−2.936 E−1	3.740 E−1	1.979 E−1	3
6.351 E−1	−9.259 E−2	2.279 E−1	1.487 E−1	6
7.996 E−1	−1.257 E−1	2.062 E−1	1.392 E−1	7
1.007 E−0	−1.165 E−1	1.761 E−1	1.249 E−1	9
1.267 E−0	−9.156 E−2	1.479 E−1	1.101 E−1	12
1.595 E−0	−3.165 E−1	1.761 E−1	1.249 E−1	9
2.009 E−0	−7.687 E−1	3.010 E−1	1.761 E−1	4
2.529 E−0				
3.183 E−0				

lived stars have ended their evolution whereas long-lived stars and inert gas add to the background. In addition, gas enriched by nucleosynthesis and returned by short-lived stars within active regions, is instantaneously mixed with unenriched gas within qiescent regions, to form a new set of identical regions for the next step. Accordingly, the ℓ-th step is characterized by k_ℓ active regions on a total of n_ℓ where, for simplifying calculations, $n_\ell = n$ shall be assumed.

The predicted oxygen abundance has a non monotonic trend in time in the sense that stars born near the end of a selected step exhibit higher oxygen abundance with respect to stars born near the beginning of the next step. In the limiting case where all regions remain quiescent, star oxygen abundance may be conceived as maintaining unchanged and the trend is represented by a continuous horizontal line on the $(\mathsf{O}\phi T)$ plane. In the limiting case where all regions are active, star oxygen abundance grows monotonically as prescribed by homogeneous simple models, regardless of the discrete (n regions) or continuous (a single region) structure of the system, and the trend is represented by a continuous inclined line on the $(\mathsf{O}\phi T)$ plane.

The special case of expected evolution, where $k_\ell = k^* = \mathrm{const}$ at any step, has been considered in an earlier attempt (CM09) for different Galactic environments, where related EDODs were fitted by models implying a strictly universal stellar initial mass function but different probabilities of a region being active. Computer runs with equal input parameters with respect to the expected evolution, shall be used in the current investigation involving stochastic star formation where, in general, $k_\ell \neq k^*$ at any step. More specifically, the following cases shall be considered: DB105 (thick disk, 2 + 2 steps); DN81 (thin disk, 5 + 20 steps); KN6 (thick + thin disk, 5 + 20 steps); H, H1 (halo, 2 + 2 steps); B (bulge, 2 + 2 steps); where the number of steps relates to the assembling and the relaxed stage, respectively. For further details refer to the parent paper (CM09, Table 7 therein, last six rows).

Which regions become active at any step, is determined by testing whether a random number generated for each region is less than a global probability deduced from the input data. Since star formation within a region is assumed to occur independently of other regions, the probability of k_ℓ active regions on a total of n_ℓ, at the cosmic time t_ℓ, is expressed by a binomial distribution. For further details refer to the parent paper (C00).

4. Results

A comparison between model predictions (TAOR, TDOD) and related data inferred from observations (EAOR, EDOD) is made with regard to the following Galactic environments: halo, bulge, thick disk, thin disk, and thick + thin disk. Two different models, H and H1, are considered for the halo as in an earlier attempt (CM09). Bulge stars are known to be old ($T \gtrsim 10$ Gyr) but large samples including age determination (to the knowledge of the author) are not available, and for this reason only model predictions are shown. The evolution of the thick + thin disk cannot be described unless the thick to thin disk mass ratio, $M_{\rm KD}/M_{\rm ND}$, is specified. As no general consensus still exists about the above mentioned mass ratio, values within the range, $0.1 \le M_{\rm KD}/M_{\rm ND} \le 0.9$, are considered. Models are fitted to oxygen abundances determined in absence of LTE approximation, and data inferred using LTE approximation are plotted only for comparison.

The number of regions is assumed to be constant for assigned stage and environment. The following values are assumed regardless of the environment: (10,10), (10,100), (100,100), (∞, ∞), where the first and the second value within brackets relate to the assembling and to the relaxed stage, respectively. The limiting case, (∞, ∞), corresponds to the expected evolution computed in an earlier attempt (CM09) and reported for comparison. For a specified case, three computer runs are performed with different sequences of random numbers for establishing whether each region is active or quiescent. The effect of stochastic star formation can be visualized by comparing related TAORs and TDODs.

With regard to R07HT subsample, a comparison between EAOR and TAOR is shown in Fig. 1 both in presence (triangles) and in absence (diamonds) of LTE approximation.

The same holds with regard to R07UT subsample both in presence (squares) and in absence (saltires) of LTE approximation. It can be seen that, in any case, the TAOR fits to the EAOR within the errors, in absence of LTE approximation, for both R07HT and R07UT subsamples. Discontinuities exhibited by TAOR imply that, at any step, the fraction of active regions, ν_ℓ, satisfies the inequality, $0 < \nu_\ell < 1, 1 \le \ell \le 4$.

With regard to R07HO and R07UO subsamples, the comparison between EDOD and TDOD is shown in Fig. 2 with same captions as in Fig. 1. It can be seen that, in any case, the TDOD fits to the EDOD within the errors, in absence of LTE approximation, for both R07HO and R07UO samples. The TDOD exhibits little or no dependence on the sequence of random numbers generated during the computer run, which implies the fraction of active regions maintains close to the probability of a region being active, $\nu_\ell \approx \chi_\ell = \chi$, at any step regardless of the total number of regions. In other words, the EDOD is very close to its counterpart related to homogeneous simple models, which is a straight line, with the

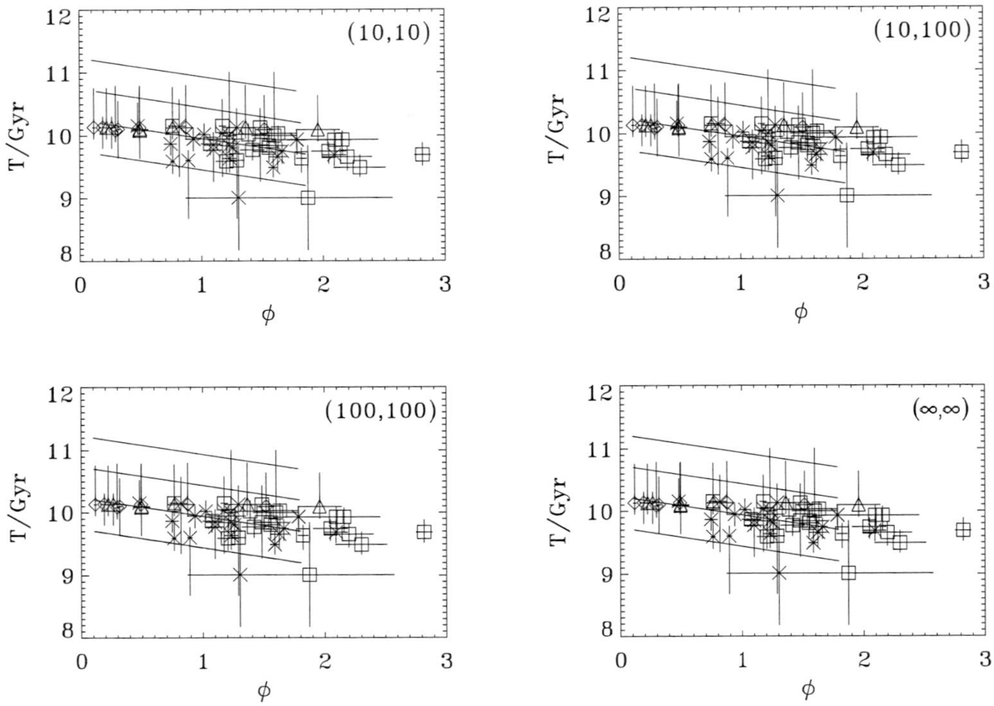

Figure 1. Comparison between theoretical (TAOR) and empirical (EAOR) age-oxygen abundance relation, with regard to halo stars (R07HT subsample, N=7) both in presence (triangles) and in absence (diamonds) of the local thermodynamical equilibrium (LTE) approximation, and stars of uncertain population (R07UT subsample, N=22) both in presence (squares) and in absence (saltires) of the LTE approximation. A cross error is superimposed to each data point. Computer runs are by model H related to three different sequences of random numbers (full, dotted and dashed curves). The number of regions at any step during the assembling (first two steps plotted from bottom left on each panel) and relaxed (remaining steps) stage are indicated on the top right of each panel. The limiting case, (∞, ∞), corresponds to the expected evolution regardless of the selected sequence of random numbers. The TAOR is independent of the beginning of evolution i.e. it can be arbitrarily shifted parallel to the vertical axis.

exception of the beginning and the end of evolution.

The EAOR plotted in Fig. 1 is compared with the TAOR calculated by use of a different model in Fig. 3, with same captions as in Fig. 1. It can be seen that all regions are active concerning at least one step, which implies no discontinuity when passing to the next step, for sufficiently low ($n = 10$) number of regions. If otherwise, $0 < \nu_\ell < 1$, $1 < \ell < 4$, and continuity never occurs. In the most extreme case (top left panel), the continuity is shown passing from the third to the fourth step (full), from the first to the third step (dashed), and from the from the first to the fourth step with the end out of scale (dotted). Further inspection shows that the TAOR fits to the EAOR within the errors provided no continuity occurs, in

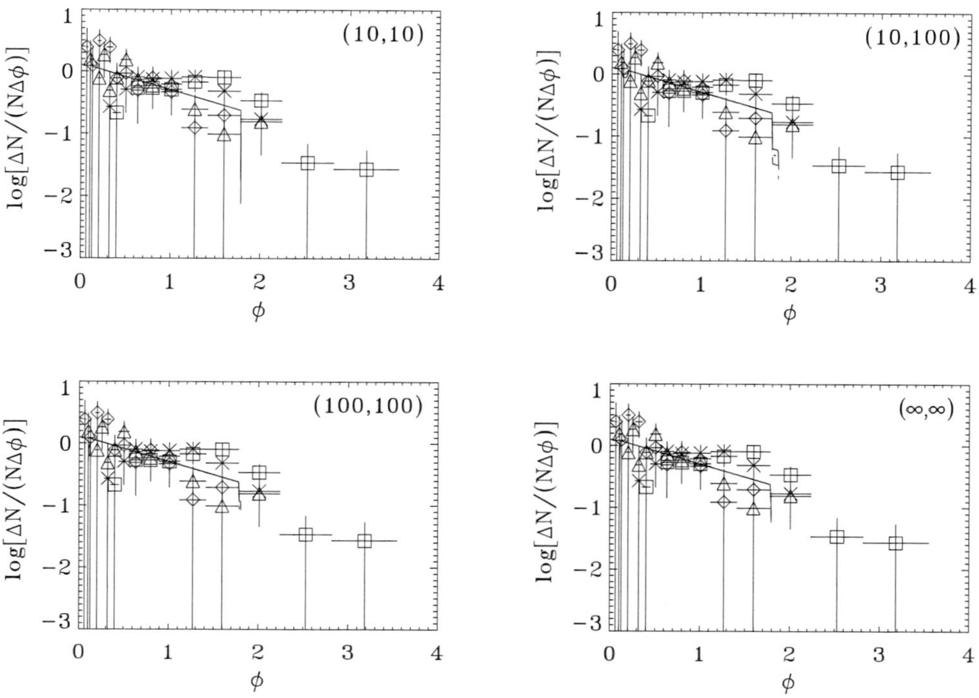

Figure 2. Comparison between theoretical (TDOD) and empirical (EDOD) differential oxygen abundance distribution, with regard to halo stars (R07HO subsample, N=28) both in presence (triangles) and in absence (diamonds) of the local thermodynamical equilibrium (LTE) approximation, and stars of uncertain population (R07UO subsample, N=51) both in presence (squares) and in absence (saltires) of the LTE approximation. A cross error is superimposed to each data point. Computer runs are by model H related to three different sequences of random numbers (full, dotted and dashed curves). The number of regions at any step during the assembling (first two steps) and relaxed (remaining steps) stage are indicated on the top right of each panel. The limiting case, (∞, ∞), corresponds to the expected evolution regardless of the selected sequence of random numbers.

absence of LTE approximation, for both R07HT and R07UT sample. If otherwise, younger low-metallicity stars could not be fitted.

The EDOD plotted in Fig. 2 is compared with the TDOD calculated by use of a different model in Fig. 4, with same captions as in Fig. 2. It can be seen that no fit is provided in any case, as outlined in the parent paper (CM09) for the expected evolution. More specifically, the EDOD deduced from a different halo sample is fitted, where oxygen abundance is deduced from an empirical [O/H]-[Fe/H] relation instead of being indirectly measured. For further details refer to the parent papers (C01; C07; CM09). Regardless of the fit, it is worth noticing that different TDODs can occur for sufficiently low number of regions ($n = 10$).

For bulge stars, no sample is available (to the knowledge of the author) where both ages and oxygen abundances have been determined, but the lack of stars younger than about 10 Gyr has been inferred (e.g., Zoccali et al., 2008). The TAOR predicted by use of a selected

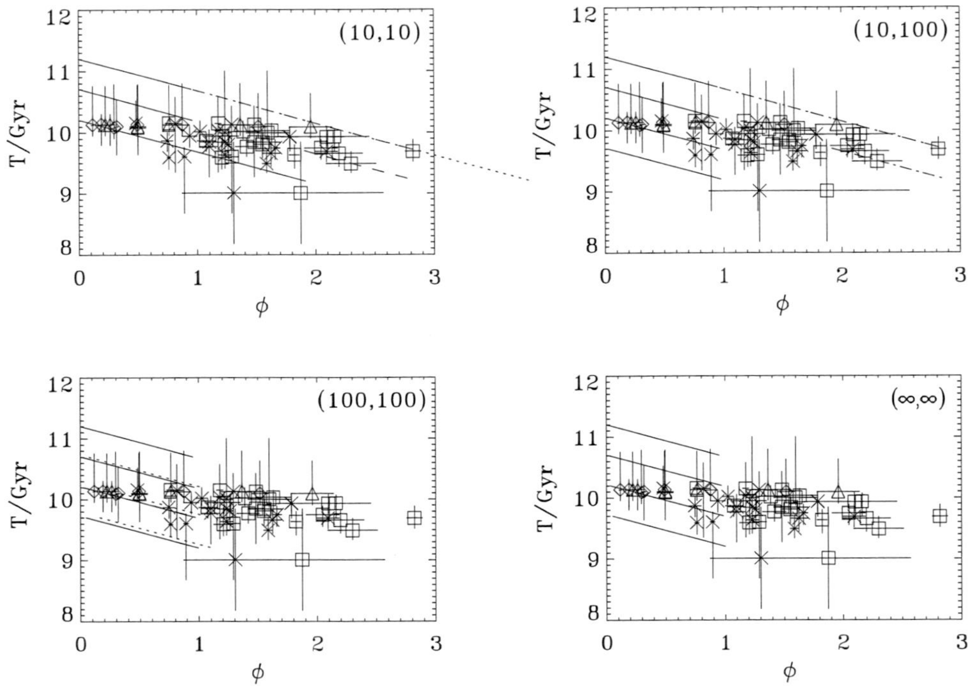

Figure 3. Same as in Fig. 1 but with computer runs by model H1. The dotted line on the top left panel ends out of scale.

model is shown in Fig. 5, with same captions as in Fig. 1 but different oxygen abundance scale. The trend is similar to what shown in Fig. 3. In particular, stochastic effects (i.e. dependence on the sequence of random numbers generated during the computer run) are relevant only for sufficiently low number of regions ($n = 10$) where the final oxygen abundance can change by a factor up to about 2 for different sequences of random numbers. With regard to M08BO subsample, the comparison between EDOD and TDOD is shown in Fig. 6 with same captions as in Fig. 2 but only in presence of LTE approximation. It can be seen that stochastic effects are relevant only for sufficiently low number of regions ($n = 10$) at the end of evolution i.e. supersolar oxygen abundances.

With regard to R07KT subsample, the comparison between EAOR and TAOR is shown in Fig. 7 both in presence (triangles) and in absence (diamonds) of LTE approximation. Captions are as in Fig. 1. It can be seen that, in any case, the TAOR provides acceptable fits to the EAOR within the errors, in absence of LTE approximation, for both R07KT and R07UT subsamples. Stochastic effects appear to be negligible.

With regard to R07KO and R07UO subsamples, the comparison between EDOD and TDOD is shown in Fig. 8 with same captions as in Fig. 2. It can be seen that, in any case, the TDOD provides acceptable fits to the EDOD within the errors, in absence of LTE approximation, for the R07KO subsample. Stochastic effects appear to be negligible, except near the beginning and the end of evolution.

With regard to R07NT subsample, the comparison between EAOR and TAOR is shown

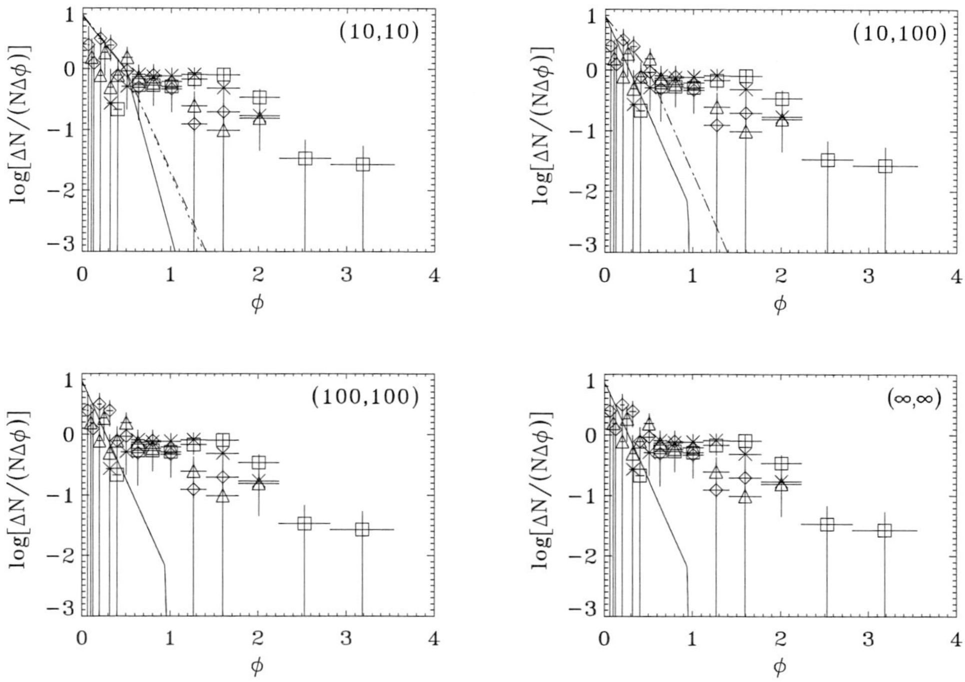

Figure 4. Same as in Fig. 2 but with computer runs by model H1.

in Fig. 9 both in presence (triangles) and in absence (diamonds) of LTE approximation. Captions are as in Fig. 1. It can be seen that, in any case, the TAOR fits to the EAOR within the errors to an acceptable extent, in absence of LTE approximation. Stochastic effects appear to be negligible unless the number of regions is sufficiently low ($n = 10$). Restricting to the R07NT subsample in absence of LTE approximation, the whole TAOR is plotted in Fig. 10. It is apparent that old metal-rich stars lie outside model predictions even if stochastic effects reduce this discrepancy.

With regard to R07NO subsample, the comparison between EDOD and TDOD is shown in Fig. 11 with same captions as in Fig. 2. It can be seen that, in any case, the TDOD provides acceptable fits to the EDOD within the errors, in absence of LTE approximation, for the R07NO subsample. Stochastic effects appear to be negligible, except near the end of evolution.

With regard to R07KT and R07NT subsamples, the comparison between EAOR and TAOR is shown in Fig. 12 (saltires and diamonds, respectively) in absence of LTE approximation. Captions are as in Fig. 1. It can be seen that, in any case, the TAOR fits to the EAOR for both R07KT and R07NT subsamples. Stochastic effects appear to be negligible unless the number of regions is sufficiently low ($n = 10$). The whole TAOR is plotted in Fig. 13. It is apparent that old metal-rich stars lie inside model predictions regardless of stochastic effects.

With regard to R07KO + R07NO subsample, the comparison between EDOD and TDOD is shown in Fig. 14 in absence of LTE approximation, where the data are normalized

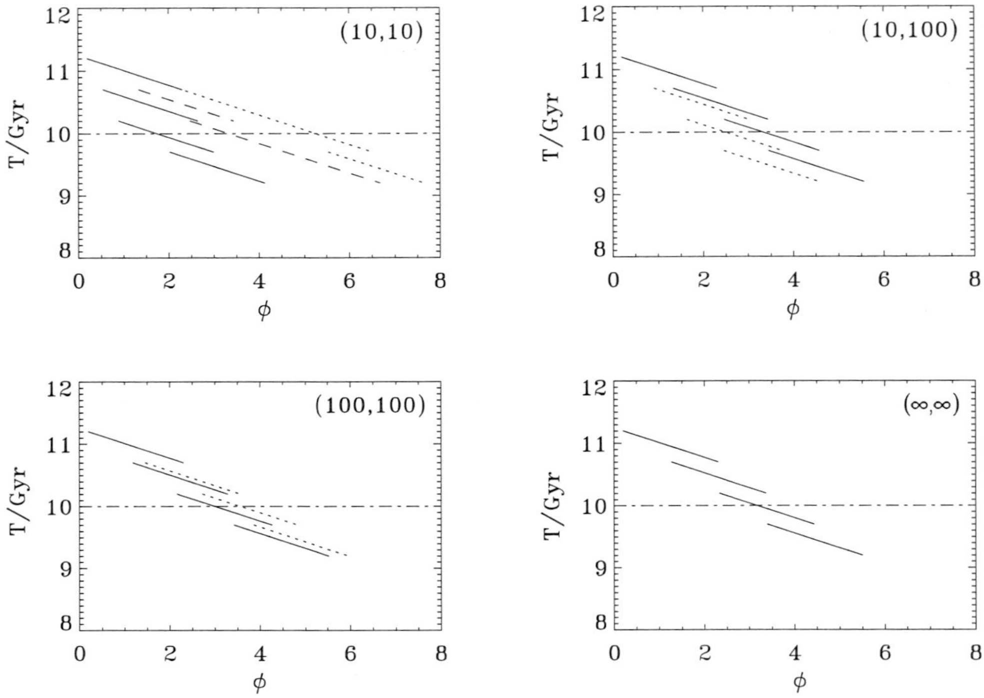

Figure 5. Theoretical (TAOR) age-oxygen abundance relation with computer runs by model B. The lack of stars younger than about 10 Gyr (below the dot-dashed line) has been inferred from observations. Other captions as in Fig. 1.

to a thick to thin disk mass ratio, $M_{\rm KD}/M_{\rm ND} = 0.1, 0.3, 0.5, 0.7, 0.9$, as represented by different horizontal error bars on the same abscissa. Captions are as in Fig. 2. It can be seen that, in any case, the TDOD provides acceptable fits to the EDOD within the errors, with the exception of higher-metallicity data (a single star) where a sole computer run makes agreement for low number of regions, case (10,10). Stochastic effects appear to be negligible, except near the end of evolution for low number of regions, case (10,10).

5. Discussion

The simple model of chemical evolution, used in the current attempt, allows both stochastic star formation and inhomogeneous mixing. More specifically, a number of k regions on a total of n is active at the beginning of each step, following a binomial distribution (C00; C07). Accordingly, different computer runs yield different TAORs and different TDODs, where related dispersions are inversely proportional to n and tend to 0 as $n \to +\infty$. In other words, stochastic effects arise for a sufficiently low number of regions, $n = 10$ in the cases considered.

At the end of any step, the remaining gas is instantaneously mixed and subdivided into n discrete regions. Gas mixing could idealize merger or accretion events in a ΛCDM

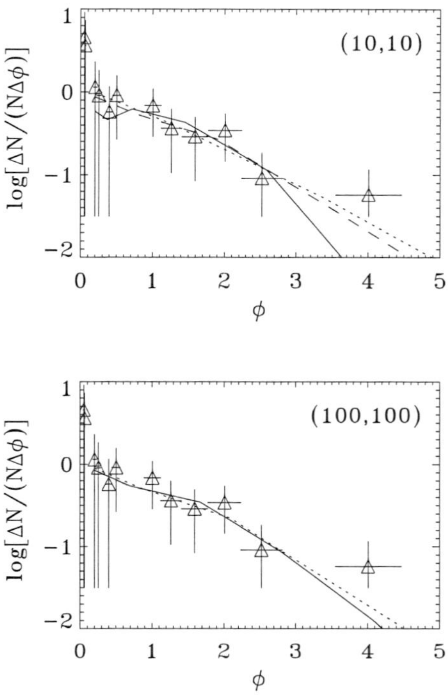

Figure 6. Comparison between theoretical (TDOD) and empirical (EDOD) differential oxygen abundance distribution, with regard to bulge stars (M08BO subsample, $N = 19$), in presence of the local thermodynamic equilibrium (LTE) approximation. Computer runs are from model B. Other captions as in Fig. 2.

cosmological scenario. Models where the number of regions is low ($n = 10$) and stochastic effects are large, could be related to discrete accretion i.e. major mergers. Conversely, models where the number of regions is high ($n \geq 100$) and stochastic effects are small, could be related to continuous accretion i.e. minor mergers or secondary infall.

The comparison between (i) EAMR and TAMR, and (ii) EDOD and TDOD, has to be considered only in a qualitative picture, owing to biases and incompleteness of the subsamples used. More specifically, R07KT-O and R07NT-O subsamples are biased towards low metallicities, [Fe/H] $\lesssim -1$ (Ral07) or $\phi \lesssim 0.3$ (CM09). The related effect is negligible to a first extent provided the low-metallicity star mass fraction within the thick and thin disk is considerably small with respect to unity, as expected. In addition, R07HT-O subsamples are incomplete and the fits shown in Figs. 1 and 2 should be confirmed using complete samples of halo stars. Finally, R07UT-O subsamples appear to be largely made of stars belonging to the thin disk (CM09). Turning to bulge stars, M08BO subsample is poorly populated and the fit shown in Fig. 6 should be confirmad using a well populated sample. On the other hand, the TDOD plotted in Fig. 6 is consistent with the EDOD deduced from a larger sample (Sadler et al. 1996) using fiducial [O/H]-[Fe/H] relations. For further details refer to the parent paper (C07).

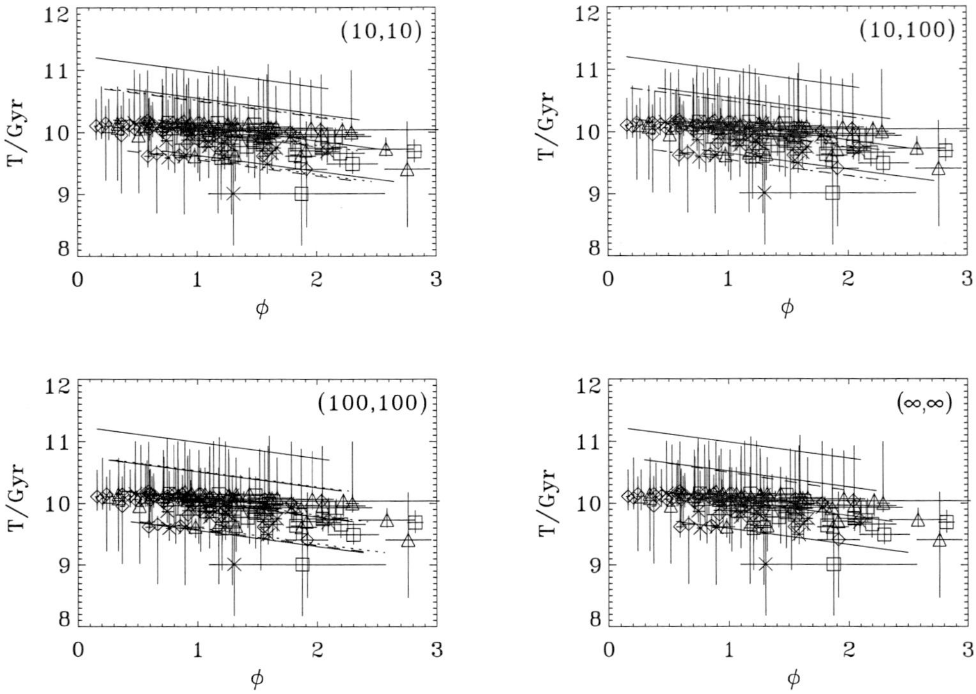

Figure 7. Comparison between theoretical (TAOR) and empirical (EAOR) age-oxygen abundance relation, with regard to thick disk stars (R07KT subsample, $N = 48$) both in presence (triangles) and in absence (diamonds) of the local thermodynamic equilibrium (LTE) approximation, and stars of uncertain population (R07UT subsample, $N = 22$) both in presence (squares) and in absence (saltires) of the LTE approximation. Computer runs are from model DB105. Other captions as in Fig. 1.

Though sample objects considered in the current paper, with the exception of bulge stars, belong to the solar neighbourhood, still related informations provide valuable clues for understanding the formation process of the Galaxy. The assumption that the EAOD and the EDOD of the local disk are representative of the global disk, can be considered as a useful zeroth order approximation. The star formation rate inferred for nearby stars is a measure of the global Milky Way star formation rate, due to radial mixing (e.g., Meusinger et al. 1991; Wielen et al. 1996; Rocha-Pinto et al. 2000; Sellwood and Binney 2002; Schönrich and Binney 2008; Roskar et al. 2008).

As outlined in an earlier attempt (Wang and Silk 1993), radial flows of the gas (possibly due to the observed spiral density waves) or different disk star formation histories (with time scale comparable to that of chemical evolution) between the inner and the outer part of the Galaxy, may change the local abundances, but the overall abundance in the disk should not be affected as long as there is no gain or loss of material in the disk. Stellar migration across the disk can be described to some extent (CM09). It can also be noticed (Wang and Silk 1993) that the average oxygen abundance in the disk is roughly the solar value, and can

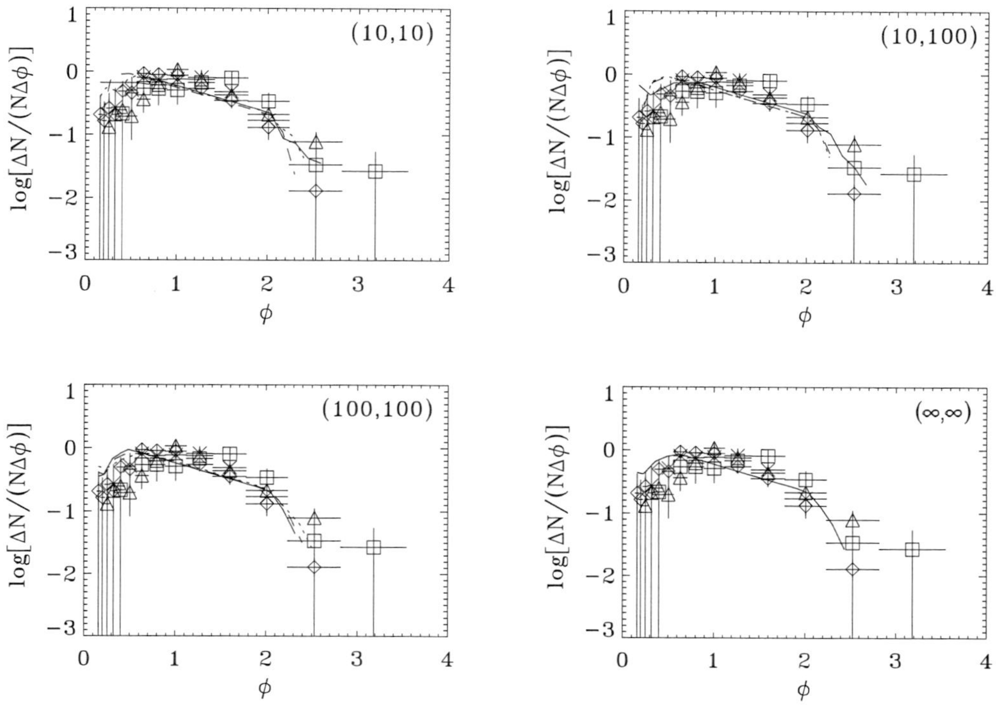

Figure 8. Comparison between theoretical (TDOD) and empirical (EDOD) differential oxygen abundance distribution, with regard to thick disk stars (R07KO subsample, $N = 133$) both in presence (triangles) and in absence (diamonds) of the local thermodynamic equilibrium (LTE) approximation, and stars of uncertain population (R07UO subsample, $N = 51$) both in presence (squares) and in absence (saltires) of the LTE approximation. Computer runs are from model DB105. Other captions as in Fig. 2.

be plausibly explained by standard (Scalo or power-law) stellar initial mass functions.

According to earlier results (CM09), the effects of stochastic star formation on both TAOR and TDOD are analysed using simple inhomogeneous models with equal stellar initial mass function but different probability of a region being active, with the exception of model H1 which is considered only for comparison. The assumption of universal stellar initial mass function implies the enhanchement of star formation via tidal interactions during merger events (Rieke et al. 1980; Doane and Mathews 1993; Lopez-Sanchez 2010) is due to a larger efficiency in the star formation rate instead of a top-heavy stellar initial mass function.

Instantaneous recycling is also assumed but the gas cannot be conceived as instantaneouly mixed, owing to the presence of active regions where star formation and evolution is going on, and quiescent regions which remain entirely gaseous. The related TAOR is necessarily discontinuous unless all regions are active, as shown in Figs. 1, 3, 5, 7, 9, 10, 12, and 13. Then the assumption of instantaneous recycling appears to affect chemical evolution to a lesser extent with respect to the assumption of instantaneous mixing, unless the

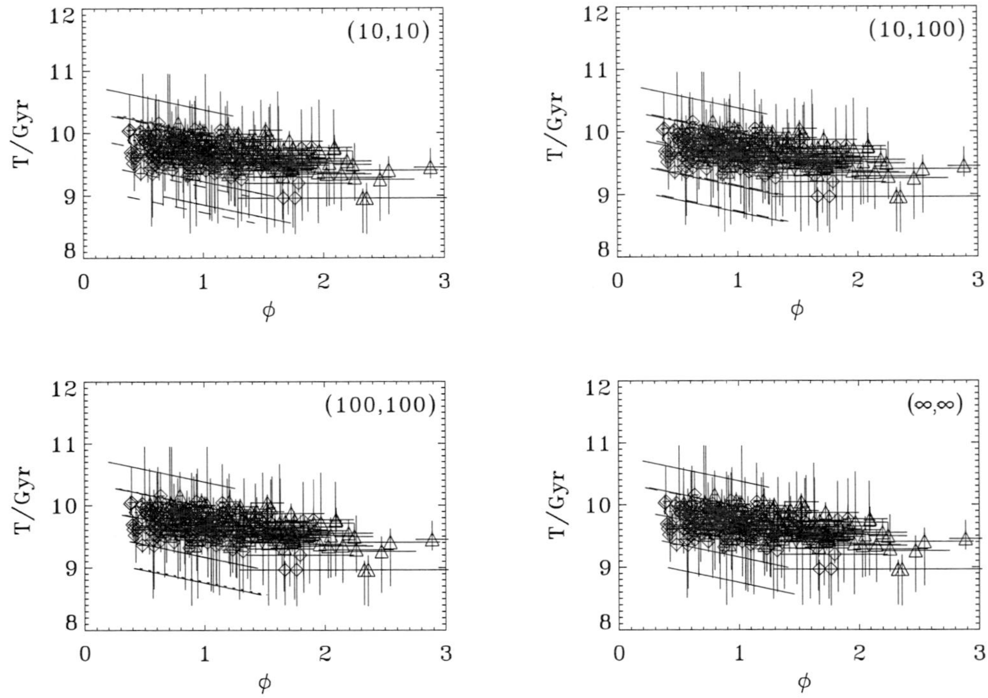

Figure 9. Comparison between theoretical (TAOR) and empirical (EAOR) age-oxygen abundance relation, with regard to thin disk stars (R07NT subsample, $N = 145$) both in presence (triangles) and in absence (diamonds) of the local thermodynamic equilibrium (LTE) approximation. Computer runs are from model DN81 where the assembling stage amounts to the first five steps, plotted from bottom left on each panel. Other captions as in Fig. 1.

gas mass fraction is extremely low.

The models of the kind considered, regardless of the stellar initial mass function, are independent of both the beginning and the end of evolution for a fixed number of steps. Accordingly, the TAOR shown in Figs. 1, 3, 5, 7, 9, 10, 12, and 13 can rigidly be translated parallel to the vertical axis, with respect to the EAOR, if a better fit is desired.

According to earlier results (CM09), the probability of a region being active is $\chi \approx 0.89$ for the B model (related to the bulge); $\chi \approx 0.32$ for the DB105 model (related to the thick disk); $\chi \approx 0.13$ for the DN81 model (related to the thin disk); $\chi \approx 0.07$ for the KN6 model (related to the thick + thin disk); $\chi \approx 0.01$ for the H model (related to the halo). The above mentioned models need neither gas outflow (or inert gas) nor gas inflow (CM09).

As shown in Figs. 1-14, stochastic effects are appreciable for sufficiently low number of regions ($n = 10$) and/or sufficiently high probability of a region being active ($\chi \gtrsim 0.8$), while they are negligible for sufficiently high number of regions ($n \geq 100$) and/or sufficiently low probability of a region being active ($\chi \lesssim 0.05$). For sufficiently high probability of a region being active ($\chi \gtrsim 0.8$), all regions can be active during one or more

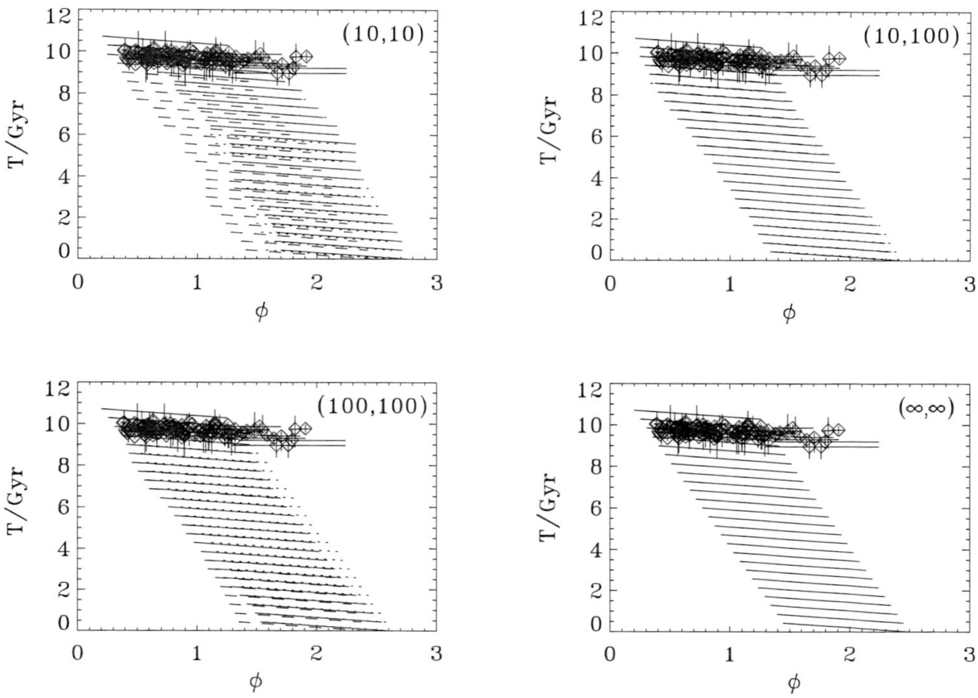

Figure 10. Same as in Fig. 9 but only in absence (diamonds) of the local thermodynamic equilibrium (LTE) approximation. The theoretical (TAOR) age-oxygen abundance relation spans over the whole evolution. Other captions as in Fig. 9.

steps, which makes the related TAOR continuous instead of discontinuous, as shown in Figs. 3 and 5. To this respect, it is worth remembering (CM09) that $\chi \approx 0.96$ for the H1 model (related to the halo but needing a large amount of gas outflow i.e. inert gas).

If the EDODs inferred from R07HO and M08BO subsamples are representative of the inner halo and the bulge, respectively, the age of halo sample stars appears to be not exceeding (within the errors) the age deduced for bulge stars, which implies a vertical shift of the TAOR related to the B model by 1 Gyr at least i.e. the onset of bulge formation at about 12.5 Gyr ago at least. A sufficiently large sample of bulge stars where ages have been determined, should be needed to get more insight on this point.

As shown in Figs. 9, 10, and 12, 13, the TAOR fits better to the EAOR, with regard to the thin disk, using a model (KN6) where the thick and the thin disk evolve as a single system, instead of a model (DN81) where the thin disk evolves independently of the thick disk. As shown in Figs. 11 and 14, the TDOD fits to the EDOD, with regard to the thin disk in absence of LTE approximation, to an acceptable extent using both KN6 and DN81 models, provided a single star related to the most oxygen-rich bin ($\phi \approx 3.2$) is conceived as an outlier and then excluded from further consideration. A fit takes place only for the TDOD determined from a single computer run where the number of regions is low ($n = 10$), with regard to KN6 model, as shown in Fig. 14.

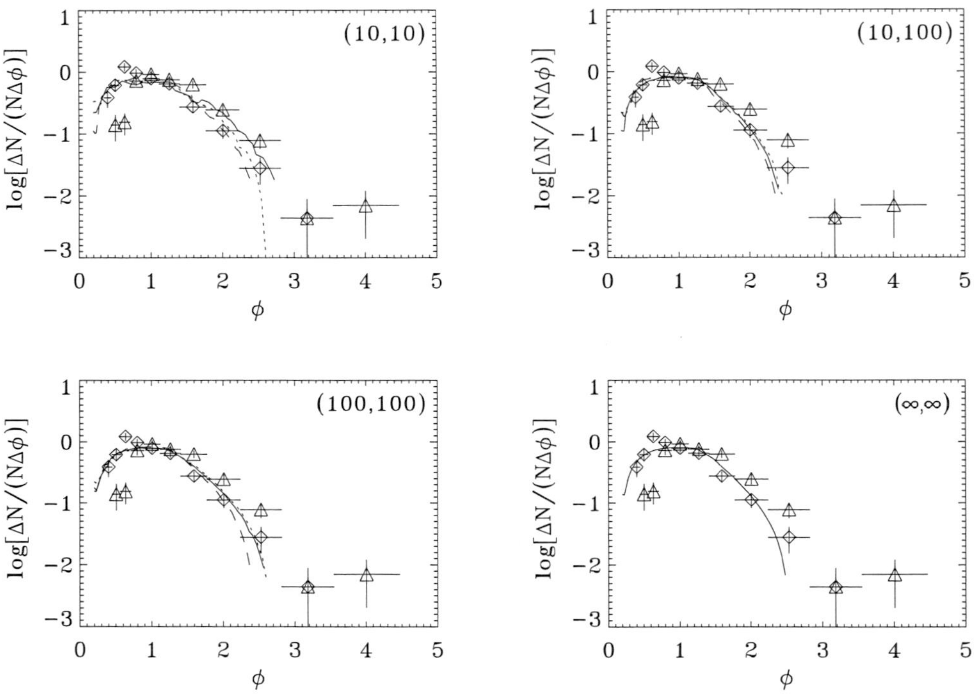

Figure 11. Comparison between theoretical (TDOD) and empirical (EDOD) differential oxygen abundance distribution, with regard to thin disk stars (R07NO subsample, $N = 310$) both in presence (triangles) and in absence (diamonds) of the local thermodynamic equilibrium (LTE) approximation. Computer runs are from model DN81, where the assembling stage amounts to the first five steps. Other captions as in Fig. 2.

The building blocks of the Galaxy and related subsystems may be conceived in a twofold manner, as lower mass overdensities which first decouple from the universe (progenitors of compact dwarf galaxies) from the cosmological side, and upper mass gas condensations which form stars (giant molecular clouds or proto-globular clusters) from the astrophysical side. Different environments are expected to imply different probabilities of a region being active. Accordingly, chemical evolution could be influenced by dynamical evolution: for instance, higher probabilities of a region being active correspond to environments where tidal effects between neighbourhing regions are strong (e.g., the bulge) and vice versa (e.g., the halo) for equal stellar initial mass function in absence of both gas outflow and gas inflow. In this view, the effects of stochastic star formation on both the TAOR and the TDOD must necessarily be considered, expecially for low number of regions ($n = 10$) and/or high probability of a region being active ($\chi \stackrel{>}{\sim} 0.8$).

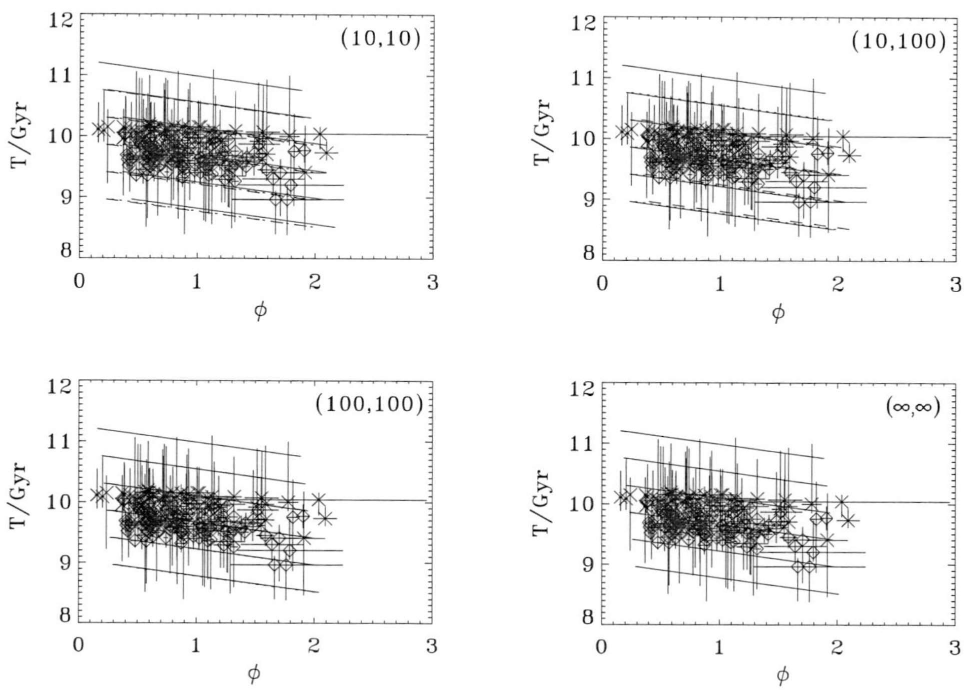

Figure 12. Comparison between theoretical (TAOR) and empirical (EAOR) age-oxygen abundance relation, with regard to thick disk stars (R07KT subsample, $N = 48$, saltires) and thin disk stars (R07NT subsample, $N = 145$, diamonds) in absence of the local thermodynamic equilibrium (LTE) approximation. Computer runs are from model KN6 where the assembling stage amounts to the first five steps, plotted from bottom left on each panel. Other captions as in Fig. 1.

6. Conclusion

A large dispersion observed in the age-metallicity relation of different Galactic environments within the solar neighbourhood, can be interpreted as due to sample selection for metallicity (Garnett and Kobulnicky, 2000) and biases in isochrone dating methods (Pont and Eyer, 2004) or in atmospheric parameters (Haywood, 2006). On the other hand, current ΛCDM cosmological scenarios imply bottom-up galaxy formation process with the occurrence of accretion and merging events where star formation is enhanced with respect to secondary infall of low-density gas. In this view, coeval stars may exhibit different metal abundance i.e. substantial intrinsic scatter. If the dispersion in the age-metallicity relation is largely due to intrinsic scatter, a natural interpretation is provided by inhomogeneous simple models of chemical evolution (Malinie et al., 1993), where the interstellar medium is structured into discrete regions and star formation occurs stochastically i.e. with assigned probability within each region (C00; C01).

The current attempt extends the results of an earlier investigation (CM09), limited to the

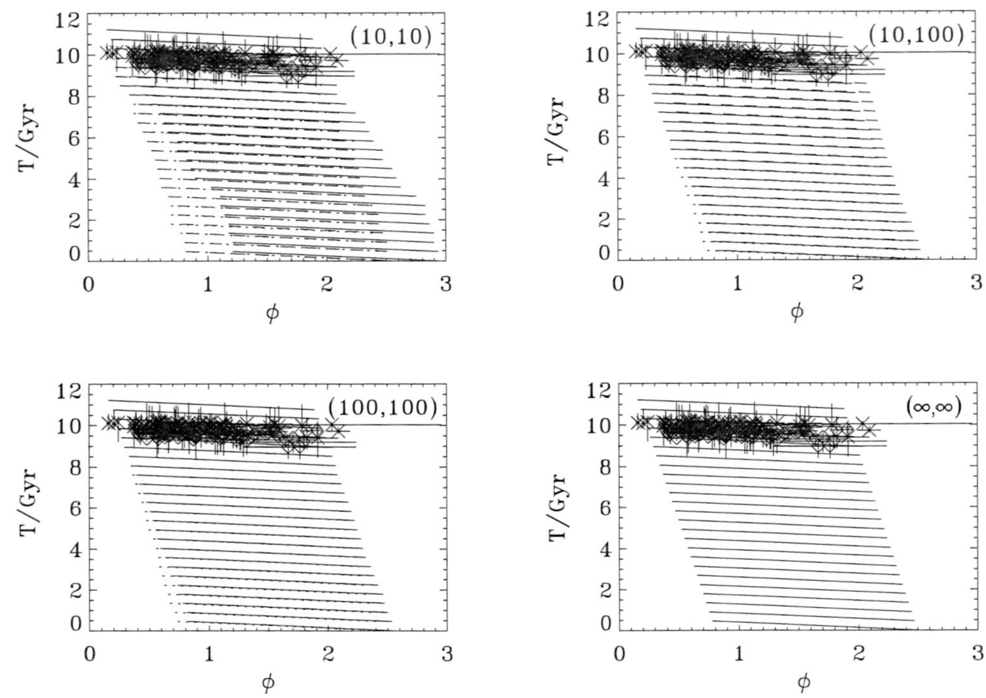

Figure 13. Same as in Fig. 12 where the theoretical (TAOR) age-oxygen abundance relation spans over the whole evolution. Other captions as in Fig. 12.

expected evolution i.e. infinitely large number of regions, to a generic stochastic star formation i.e. finite number of regions. The selected models are characterized by the following features (CM09): (i) an initial assembling stage where the volume undergoes substantial change [2 steps for the halo (H), the bulge (B), and the thick disk (DB105); and 5 steps for the thin disk (DN81) and the thick + thin disk (KN6)] and a subsequent relaxed stage where the volume remains more or less unaltered (2 steps for H, B, and DB105 models; and 20 steps for DN81 and KN6 models); (ii) initial and final oxygen abundance inferred from related subsamples; (iii) universal stellar initial mass function both in space (different environments) and in time (different stages) with neither gas outflow nor gas inflow; (iv) same effective yield for H, DB105, KN6 models; about one half larger for DN81 model; about three quarters larger for B model; (v) the theoretical differential oxygen abundance distribution (TDOD) fits to the empirical differential oxygen abundance distribution (EDOD) to an acceptable extent in all cases, in absence of LTE approximation, provided the higher oxygen-abundance bin (related to a single star) is disregarded as biased or outlier.

The main results of the current attempt may be summarized as follows: (1) the theoretical age-oxygen abundance relation (TAOR) predicted by the model fits to the empirical age-oxygen abundance relation (EAOR) to an acceptable extent (concerning at least one computer run) for both the halo, the thick disk, and the thick + thin disk regardless of the number of regions, while the same holds for the thin disk only if the number of regions is

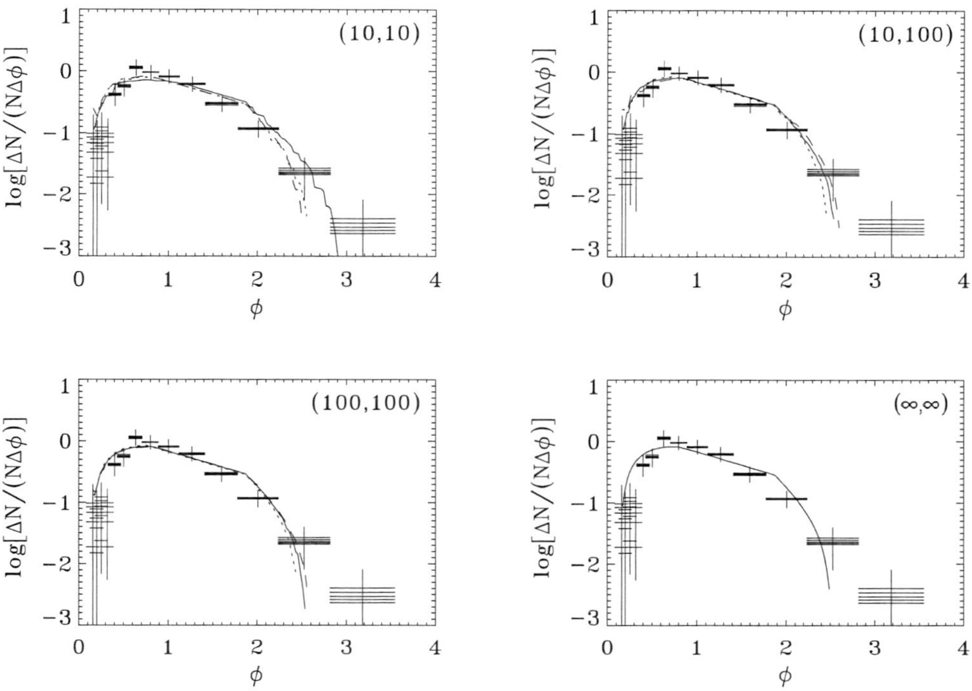

Figure 14. Comparison between theoretical (TDOD) and empirical (EDOD) differential oxygen abundance distribution, with regard to thick + thin disk stars (R07KO subsample, $N = 133$; R07NO subsample, $N = 310$), normalized to a thick to thin disk mass ratio, $M_{\rm KD}/M_{\rm ND} = 0.1$, 0.3, 0.5, 0.7, 0.9, as represented by different horizontal error bars on the same abscissa, in absence of the local thermodynamic equilibrium (LTE) approximation. Computer runs are from model KN6, where the assembling stage amounts to the first five steps. Other captions as in Fig. 2.

low ($n = 10$) i.e. owing to stochastic effects; (2) the TDOD predicted by the model fits to the EDOD to an acceptable extent (concerning at least one computer run) for both the halo, the bulge, the thick disk, the thin disk, and the thick + thin disk, regardless of the number of regions, while the fit is further improved for the thick + thin disk if the number of regions is low ($n = 10$) i.e. owing to stochastic effects.

The above results need to be confirmed using higher-population (bulge, halo), complete (halo), unbiased towards low abundances (disk) samples, which are not (to the knowledge of the author) presently available. If subsamples considered in the current investigation are representative of related environments, stochastic effects due to inhomogeneous star formation make a viable interpretation of the large scatter exhibited by the EAOR in different stellar populations.

Acknowledgements

Thanks are due to S. Rubele for performing computer runs and drawing a preliminary version of a large fraction of the figures.

References

[1] Brosche, P. *Astron. Astrophys.* 1970, 6, 240-253.

[2] Caimmi, R. *Astron. Nachr.* 2000, 321, 323-362 (C00).

[3] Caimmi, R. *Astron. Nachr.* 2001a, 322, 65-69 (C00, erratum).

[4] Caimmi, R. *Astron. Nachr.* 2001b, 322, 241-263 (C01).

[5] Caimmi, R. *New Astron.* 2007, 12, 289-321 (C07).

[6] Caimmi, R. *New Astron.* 2008, 13, 314-339.

[7] Caimmi, R.; Milanese, E. *Astrophys. Space Sci.* 2009, 323, 147-167 (CM09).

[8] Cartledge, S.I.B.; Lauroesch, J.T.; Meyer, D.M.; Sofia, U.J. *Astrophys. J.* 2006, 641, 327-346.

[9] Doane, J.S.; Mathews, W.G. *Astrophys. J.* 1993, 419, 573-584.

[10] Edvardsson, B.; Andersen, J.; Gustaffson, B.; et al. *Astron. Astrophys.* 1993, 275, 101-152.

[11] Garnett, D.R.; Kobulnicky, H.A. *Astrophys. J.* 2000, 532, 1192-1196.

[12] Haywood, M. Mon. Not. R. *Astron. Soc.* 2006, 371, 1760-1776.

[13] Lin, C.C.; Shu, F.H. *Astrophys. J.* 1964, 140, 646-655.

[14] Lin, C.C.; Yuan, C.; Shu, F.H. *Astrophys. J.* 1969, 155, 721-746.

[15] Lopez-Sanchez, A.R. *Astron. Astrophys.* 2010, 521, A63, 1-31.

[16] Malinie, G.; Hartmann, D.H.; Clayton, D.D.; Mathews, G.J. *Astrophys. J.* 1993, 413, 633-640.

[17] Melendez, J.; Asplund, M.; Alves-Brito, A.; et al. *Astron. Astrophys.* 2008, 484, L21-L25.

[18] Meusinger, H.; Stecklum, B.; Reimann, H.-G. *Astron. Astrophys.* 1991, 245, 57-74.

[19] Nittler, L.B. *Astrophys. J.* 2005, 618, 281-296.

[20] Pagel, B.E.J. In *Evolutionary Phenomena in Galaxies;* Beckman J.E.; Pagel B.E.J.; Eds.; A90-31743 13-90; Cambridge Univ. Press: Cambridge and New York, UK, 1989;, pp 201-223.

[21] Pilyugin, L.S. *Astron. Astrophys.* 1996, 313, 803-809.

[22] Pont, F.; Eyer, L. *Mon. Not. R. Astron. Soc.* 2004, 351, 487-504.

[23] Prantzos, N. *In Stellar Nucleosynthesis 50 years after B2FH*; EAS Publication Series; EDP Sciences: 2008; Vol. 32, pp.311-356.

[24] Ramirez, I.; Allende Prieto, C.; Lambert, D.L. *Astron. Astrophys.* 2007, 465, 271-289.

[25] Rieke, G.H.; Lebofsky, M.J.; Thompson, R.I.; et al. *Astrophys. J.* 1980, 238, 24-40.

[26] Rocha-Pinto, H.J.; Maciel, W.J. *Mon. Not. R. Astron. Soc.* 1996, 279, 447-458.

[27] Rocha-Pinto, H.J.; Scalo, J.M.; Maciel, W.J.; Flynn, C. *Astrophys. J.* 2000, 531, L115-L118.

[28] Roškar, R.; Debattista, V.P.; Quinn, T.R.; et al. *Astrophys. J.* 2008, 684, L79-L82.

[29] Ryan, S.G.; Norris, J.E. *Astron. J.* 1991, 101, 1865-1878.

[30] Sadler, E.M.; Rich, R.M.; Terndrup, D.M. *Astron. J.* 1996, 112, 171-185.

[31] Schönrich, R.; Binney, J. *Mon. Not. R. Astron. Soc.* 2009, 399, 1145-1156.

[32] Sellwood, J.A.; Binney, J. *Mon. Not. R. Astron. Soc.* 2002, 336, 785-796.

[33] Wang, B.; Silk, J. *Astrophys. J.* 1993, 406, 580-589.

[34] Wielen, R.; Fuchs, B.; Dettbarn, C. *Astron. Astrophys.* 1996, 314, 438-447.

[35] Wilmes, M.; Köppen, J. *Astron. Astrophys.* 1995, 294, 47-56.

[36] Zoccali, M.; Hill, V.; Lecureur, A.; et al. *Astron. Astrophys.* 2008, 486, 177-189.

In: Galaxies: Evolution, Morphology and Dynamics ISBN: 978-1-62081-185-6
Editors: A. Guerrera Sanchez et al., pp. 135-147

Chapter 5

RADIO QUIET AGN PROPERTIES VS. SPIN-PARADIGM: THE CASE OF NGC 4945

E. Fedorova * ***and V.I. Zhdanov*** †
Astronomical Observatory of Kiev National Taras Schevchenko University
Kiev, Ukraine

Abstract

The most accepted model to explain the nature of RQ/RL dichotomy of AGN is the so-called spin paradigm, which explains the jets and powerful radio emission of RQ objects as an effect of higher values of BH spin comparably to RQ ones. One of the observational differences between RQ and RL AGN predicted by spin-paradigm lies in the form of the hard X-ray primary continuum of the central machine: for the RL objects it have to be harder than for RQ ones, and have the exponential cut-off below 100 keV. However, at least several galaxies contradicting to this prescription were reported earlier, and now we report the similar properties of RQ Seyfert 2 NGC 4945, basing on INTEGRAL six-years observational data. NGC 4945 has an unstable spectral shape in hard X-rays, and phases with hard spectrum and exponential cut-off below 100 keV were observed here. Thus the necessity of further comparative analysis of the radio/optical/X-ray properties of such objects is stressed out.

PACS 98.54.-h, 98.62.Js, 98.54.Cm, 97.80.Jp. **Keywords:** Active Galactic Nucleus.

1. Introduction

1.1. The Spin Paradigm

The nature of radio-loudness(RL) or radio-quietness(RQ) of AGNs remains an open astrophysical problem today. The different attempts have been enterprized to explain it in terms of orientation likewise the geometrical scheme by Antonucci & Miller did for the type 1/type 2 AGNs [1], as well as in terms of higher masses of RL objects comparatively

*E-mail address: efedorova@ukr.net
†E-mail address: ValeryZhdanov@gmail.com

to RQ ones [8, 21]. The first class of models had been ruled out by the observational results: as we know today, the type 1 and type 2 RQ AGNs coexist with RL ones, thus the only link between radio-loudness and geometry can be traced in blazars where the jet and the line-of-sight are parallel [2, 22]. The second class of models, presuming the RL objects to be more massive, also was not confirmed by observations [31], despite the fact that the jets in more massive AGNs are more prominent than in less massive ones. However, jets can exist in the objects with whole range of the central black hole (BH) masses, i.e. from 10^6 up to 10^9 $M_\odot$ [31], thus the higher mass cannot be the main cause of the radio-loudness.

The so-called "spin-paradigm", which is today the most successful model to explain the nature of RQ/RL dichotomy of AGN, was primarily proposed by Wilson & Colbert [30] and developed by Blandford [4]. Within this model the jets and powerful radio emission of RL objects could be explained as an effect of higher values of BH spin comparably to RQ ones. If the BH spin is higher than the angular momentum of the accretion disk at the inner edge, the energy from the BH ergosphere can be extracted via Blandford-Znajek mechanism [6]. Then the poloidal magnetic field of the accreting matter forms jets from the innermost parts of the accretion disk via Blandford-Payne mechanism [5], forming the gap around the BH. Normalized spin j lies between 0 and 1 and jet power depends on j^2. As the jet power depends on the spin value it is connected directly with the radius of the inner edge of the accretion disk (or with the disk gap): the greater gap sizes are prescribed to the disks around BH with higher spin values. This causes the observational difference between RQ and RL AGN which lies in the form of the hard X-ray primary continuum of the central machine: for the RL objects it have to be harder than for RQ ones, and have the exponential cut-off below 100 keV.

Recently the spin-paradigm was reviewed by Garofalo et al. ([11]) and adopted to the case of retrograde systems (i.e. disk rotating in the direction opposite to the BH spin). This model explains the most prominent jet activity by the retrograde disk; prograde systems can have no jets even with comparably high values of the BH spin. But the gap play the same role in this model, so the prescription about the link between the inner radius of an accretion disk (and thus the exponential cut-off) and the jet power remains.

However, some observational results contradicting to this prescription for several galaxies were reported recently (NGC 4388, IC 235A, NGC 5548). Here we analyze a similar properties of the other RQ Seyfert 2 galaxy, NGC 4945, basing on INTEGRAL six-years observational data.

1.2. Compton Thick Seyfert 2 Galaxy NGC 4945

NGC4945 is the one of the brightest known Seyfert 2 galaxies, near edge-on (inclination angle is 80°), with the redshift z=0.0019 [29]. It is a spiral galaxy with quite complicate structure of the central parts, consisting of active galactic nucleus (AGN), starburst regions and conical-shaped plume, possibly caused by nuclear starburst [25]. It is also a powerful source of a megamaser/starburst emission which is very intensive both at radiowaves and at hard γ-rays [17] meanwhile its AGN is hardly visible here. Thus this AGN was identified as an RQ one [13]. Here we analyze the properties of this AGN from the INTEGRAL data obtained during the six-years monitoring of this object.

The first wide-band 0.1-200 keV BeppoSAX NGC 4945 spectrum was also obtained by Guinazzi et al.[12]; these observations were lasting 3 days. Authors found the overall spectrum of NGC 4945 consisting of three main parts:

1. reprocessed primary continuum, slightly absorbed;

2. thermal emission from the bulge region;

3. primary nuclear continuum (probably Compton-reflected), heavily absorbed (the best-fit model for this component was the power-law with slope 1.5 and cut-off above 100 keV, and absorbtion at level of $3.9 * 10^{24}$ cm^{-2}).

This third part, manifesting itself brightly at energies higher than 10 keV, is of the main interest of this work. Due to its composite nature (AGN plus starburst) it's necessary to analyze its spectrum at energies higher than 10 keV [12].

The results of the first year of observations of NGC 4945 by INTEGRAL mission (276 ks, Jan 2003 - Jan 2004) had been already presented by Soldi et al. [27]. This data were insufficient yet to disclose the cut-off at high-energies, Compton reflection or other fine properties of AGN in NGC 4945 due to comparatively low-counting statistics. They found in 20-200 keV energy range the spectral slope of 1.9 using fixed value for hydrogen column density $N_H = 4 * 10^{24}$ cm^{-2} and no essential constrictions on high-energy cut-off. Their results on BeppoSax/PDS spectrum show the presence of the high-energy cut-off $E_c = 130^{+255}_{-55}$ keV, spectral slope of 1.6±0.3 and hydrogen column density $N_H = (4.5 \pm 0.7) * 10^{24}$ cm^{-2}.

This object had been also observed several times by Suzaku [16, 19]; these observations had constrained the cut-off at high energies ($E_C = 250^{+150}_{-50}$ keV) with slope of 1.6±0.1. But the most remarkable result of Suzaku observations was recently reported by Marinucci et al.[19]: the N_H variability.

Here we confirm the variations of the absorption from the INTEGRAL data on NGC 4945, however, demonstrating also that the spectral changes in 20-500 keV observed here cannot be explained in terms of absorbtion variability only.

2. Data Processing

A summary of the *INTEGRAL* observations that are analyzed in this paper is given in Table 1. The data sample includes all publicly available *INTEGRAL* data on the source as of September 2011, i.e. 2470 ScW (science windows)[1] from spacecraft revolutions from 046 up to 776 for the IBIS/ISGRI instrument. For both JEM-X monitors there were 59 and 28 science windows, respectively. The total ISGRI exposure time of the dataset which had been used to extract the lightcurves is $\sim$ 4.2 Ms (including all the observations when the object was at the angle less than 15° off-axis). For spectral extraction the smaller dataset (from the area within the 10° off-axis angle) of 864 ScWs (total exposure $\sim$ 2.1 Ms) had been used. For the two X-ray monitors smaller set of data is available $\sim$ 300 Ks (off-axis angle $< 3°$).

INTEGRAL IBIS/ISGRI data analysis was performed with version 9.0 of the Offline Scientific Analysis software (OSA) which is provided by the ISDC [7]. For IBIS/ISGRI

[1]A science window is an *INTEGRAL* dataset based on typically 2 ks of observation.

Table 1. ***INTEGRAL*** **observation log for NGC 4945.**

Data rev	Start date [UTC]	ISGRI exp. [ks]	JEM-X1 exp. [ks]	JEM-X2 exp. [ks]
046-049	2003-03-02T09:58	153.5	0	6.9
076-092	2003-05-30T00:54	361.3	0	0
146-150	2004-12-24T08:18	4.3	0	59.5
157-197	2004-01-26T11:38	337.0	0	12.6
201-235	2004-06-05T19:16	55.7	0	0
258-284	2004-11-23T13:13	100.2	0	0
322-349	2005-06-09T01:42	973.7	0	0
350-399	2005-08-25T12:04	160.8	41.6	0
400-446	2006-01-22T09:42	213.7	116.8	1.8
455-477	2006-07-06T15:36	108.7	0	0
514-531	2006-12-29T20:19	1331.2	0	0
573-588	2007-06-25T06:37	372.5	0	0
636-648	2007-12-31T03:31	298.4	40.0	0
651-699	2008-02-11T21:25	233.1	0	0
700-719	2008-07-08T20:42	780.4	0	0
764-776	2009-01-15T20:50	415.1	13.6	0
Total	03/2003 – 08/2009	5900.6	212.0	70.9

spectral extraction the standard recipe was applied. All the spectra were extracted individually for every science window and then summed up into the whole set. The source is detected up to 500 keV. In order to account for (inter)calibration uncertainties, we applied 3% systematics to the *INTEGRAL* spectra. The significances of detection are 120.5σ (20 – 40 keV), 100.1σ (40 – 60 keV), 64.1σ (60 – 100 keV), and 31.0σ (100 – 200 keV). For the JEM-X data we also have used OSA 9 software, and we had extracted the spectra using the standard recipe for this version.

The IBIS/ISGRI lightcurves have been extracted in four energy bands (20–40 keV, 40–60 KeV, 60–100 keV and 100–200 keV). The time bin of the lightcurves varies from 1 to 5 days, depending on the exposure time of the particular *INTEGRAL* revolution (averaging was performed in every particular data revolution or, in the case of short time exposure during one revolution, in small group of two or three revolutions following one after another).

3. Spectra and Variability

Spectral modelling of the ISGRI and JEM-X spectra of NGC 4945 had been performed with the XSPEC 12 package of HEASoft 6.11 [3]. We confine ourselves by the energies higher 7 keV to analyze the primary continuum emission of this object. Thus in the spectral treatment we had considered only two following models which describe the effect of reflection of the primary AGN emission on either the accretion disk, a hot corona or the cold absorbing gas surrounding it (see e.g. [23] for a physical motivation of the choice of

Table 2. The best-fit model (*ii*) parameters to the averaged *INTEGRAL* ISGRI and combined JEM-X spectra with the errors at 90% confidence level.

Parameter	Model *i*	Model *ii*
χ^2/d.o.f	14.6/14	11.7/13
Γ	1.69 ± 0.03	1.38 ± 0.04
E_C [keV]	-	175^{+20}_{-17}
R	$5.0^{+10.0}_{-2.0}$	$0.7^{+0.4}_{-0.3}$
N_H, 10^{24} cm^{-2}	2.3 ± 0.4	2.8 ± 0.4
$F_{10-20\,\mathrm{keV}}$, $10^{-11}\,\mathrm{erg\,cm^{-2}\,s^{-1}}$	3.29 ± 0.03	3.33 ± 0.03
$F_{20-40\,\mathrm{keV}}$, $10^{-11}\,\mathrm{erg\,cm^{-2}\,s^{-1}}$	9.0 ± 0.08	8.90 ± 0.08
$F_{40-60\,\mathrm{keV}}$, $10^{-11}\,\mathrm{erg\,cm^{-2}\,s^{-1}}$	7.08 ± 0.05	7.19 ± 0.06
$F_{60-200\,\mathrm{keV}}$, $10^{-11}\,\mathrm{erg\,cm^{-2}\,s^{-1}}$	9.48 ± 0.11	9.7 ± 0.1
$F_{100-200\,\mathrm{keV}}$, $10^{-11}\,\mathrm{erg\,cm^{-2}\,s^{-1}}$	13.0 ± 0.2	11.7 ± 0.1
$F_{>200\,\mathrm{keV}}$, $10^{-11}\,\mathrm{erg\,cm^{-2}\,s^{-1}}$	14.6 ± 0.2	8.38 ± 0.08
K_X	1.36 ± 0.28	1.3 ± 0.3

the spectral models):

1. an absorbed power-law model with cold Compton reflection component (XSPEC model `zwabs·reflect(po), by Magdziarz & Zdziarski [18]);`
2. an absorbed power-law model with additional cold Compton reflection component and exponential cut-off at high energies (`zwabs·pexrav` model, [18]).

In order to account for the different observational periods and for uncertainties in the cross-calibration, we applied intercalibration factor K_X for the JEM-X spectrum with respect to the ISGRI spectrum. The fluxes in the six energy bands (10–20 keV, 20–40 keV, 40–60 keV, 60–100 keV, 100–200 keV and 200-500 keV) are given in Table 2.

The best-fit model to the overall spectra following the F-test statistics is the model *ii* (F-test probability 94% and F-value 3.22). This model gives a photon index and high-energy cut-off consistent with the one found in *BeppoSAX* observations [12] with an intrinsic absorption remarkably lower. The high-energy cut-off is also slightly lower than one obtained for this object by Itoh et al.[16] from the Suzaku observational data. The reflection parameter also can be determined from this data (Table 2), appearing to be 0.7. The folded *INTEGRAL* spectra are shown in Fig. 1.

To look into possible variability of NGC 4945 over the observational period and its origins we had extracted the lightcurves from *INTEGRAL* IBIS/ISGRI data in the 20–40 keV, 40–60 keV and 60–100 keV energy bands. The ISGRI lightcurves together are shown in Fig. 2[2]. The variances of these three lightcurves is 6.48 in 20–40 keV energy band, 6.9 in 40–60 keV band and 3.97 in 60–100 keV band. As we can see from the lightcurves there no prominent correlation between them during the hole period of observations. This can be

[2]Time is given in IJD (*INTEGRAL* Julian Date). The origin of IJD is January 1st, 2000 expressed in terrestrial time.

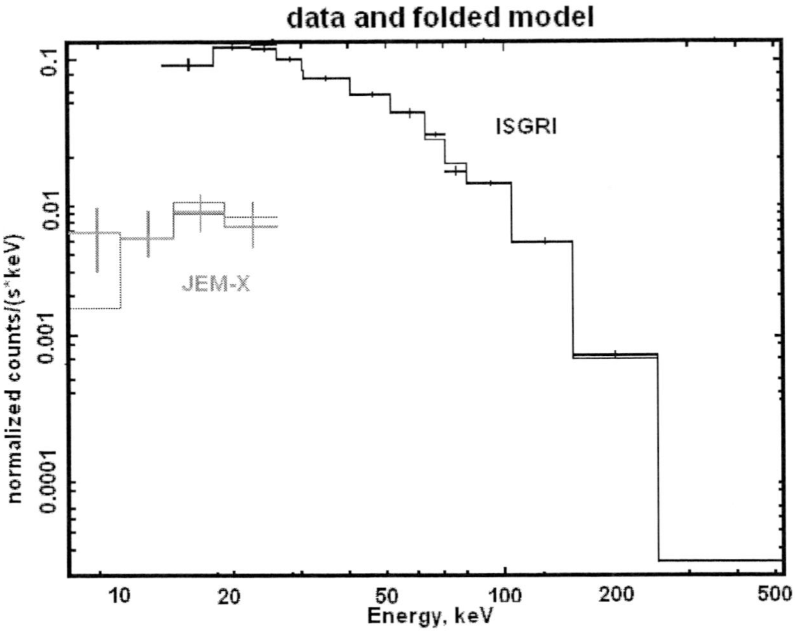

Figure 1. Total time ISGRI and JEM-X spectra.

the first sign of the spectral variability which is more complex than it was disclosed from *Suzaku* observations [19], i.e. not only N_H are varying with time. This motivated us to investigate the spectral parameters evolution with time.

Thus we fitted separately the IBIS/ISGRI spectra corresponding to each revolution. To make our datasets as uniform as possible we did not consider here the JEM-X data, which cover only a limited part of the ISGRI observations. As the statistics in every particular data revolution set is too low to see fine spectral features, to fit these data and to estimate the fluxes in several energy bands, the absorbed `pexrav` model with the reflection parameter frozen to the value of 0.7 obtained to the total-time spectrum was applied.

To analyze the spectral shape changes several models based on the best-fit to the total exposure time spectrum were tested:

1. model with E_C and Γ fixed to the mean values (only N_H varies) was used as a "null-hypothesis";
2. "two-parametric" model with variable N_H and photon index;
3. "two-parametric" model with variable N_H and E_C;
4. all the three parameters N_H, E_C and Γ are free to vary.

We also had tested the models with changeable reflection parameter and, in the major part of the data, the only result for it is an upper limit on its value, compatible with one obtained for the total exposure time spectrum. We do not mean here the variants with fixed absorbtion value as physically non motivated.

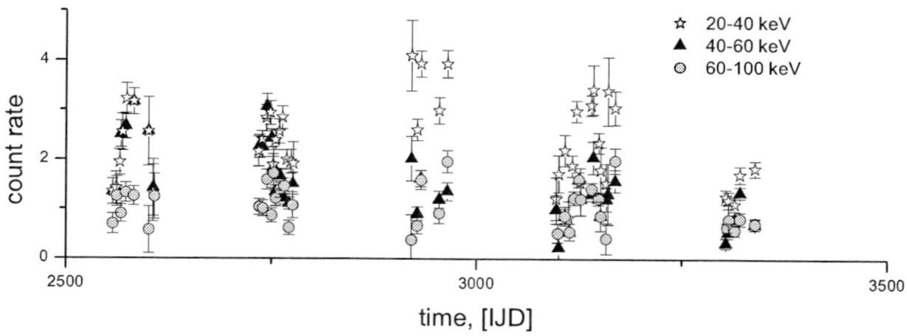

Figure 2. The best-quality fragment of the ISGRI lightcurves of NGC 4945 in three energy bands (during 2006–2009).

The first model appears to be no applicable due to the high values of reduced chi-squared it give in more than a half of the data revolution spectra (above 2.0). The other two fits are better but for significant part of the spectra is not comparable with three-parameter fit yet. The dependencies of best-fit parameters from time are shown on the Fig.3.

The variances of the parameters shown on the Fig.3 are 35.93 (HECs), 2.56 (photon indices) and 1.43 (for hydrogen column density), confirming that all the three of them are really varying. To trace out the link between these parameters, we had calculated the correlation between them. As the result, we have quite high correlation between spectral slope and high-energy cut-off, r=0.76 (significant positive correlation). Correlation between hydrogen column density and high-energy cut-off is 0.08 (no correlation), as well as between hydrogen column density and spectral slope it is 0.14 (correlation is absent or extremely weak).

As we can see from the Fig.3, all the existing data on NGC 4945 can be divided in parts following the next criteria:

1. periods of low absorbtion vs. periods of high one (comparably to the mean value);

2. periods of steep spectrum vs. flat one, or periods with high-energy cut-off below 100 keV (as it is typically for radio-loud AGNs), periods with cut-off above 100 keV, and periods with no cut-off (or higher than 500 keV).

If we apply the first criteria, we will have the "opaque phase" (OP) and "transparent phase" (TP); for the second one we can say the "steep spectrum phase" and flat spectrum phase, or "lo-cut" (cut-off below 100 keV), "hi-cut" (above 100 keV) and "no-cut" (no cut-off/above 500 keV) phases. The spectral fitting parameters within the model *ii* for OP, TP, "lo-cut", "hi-cut" and "no-cut" phases are shown in the Table 3.

The differences between N_H values during the OP/TP phases, as well as the N_H variance confirm the variability of it disclosed by Mariucci et al.[19] from *Suzaku* observations.

Cut-off values of $E_C \simeq 140\,\mathrm{keV}$ and $E_C \simeq 73\,\mathrm{keV}$ are measured during the "hi-cut" and "lo-cut" phases, respectively, and only a lower limit of $E_C > 580\,\mathrm{keV}$ is found during the "no-cut" phase. The photon indices during these phases show significant differences,

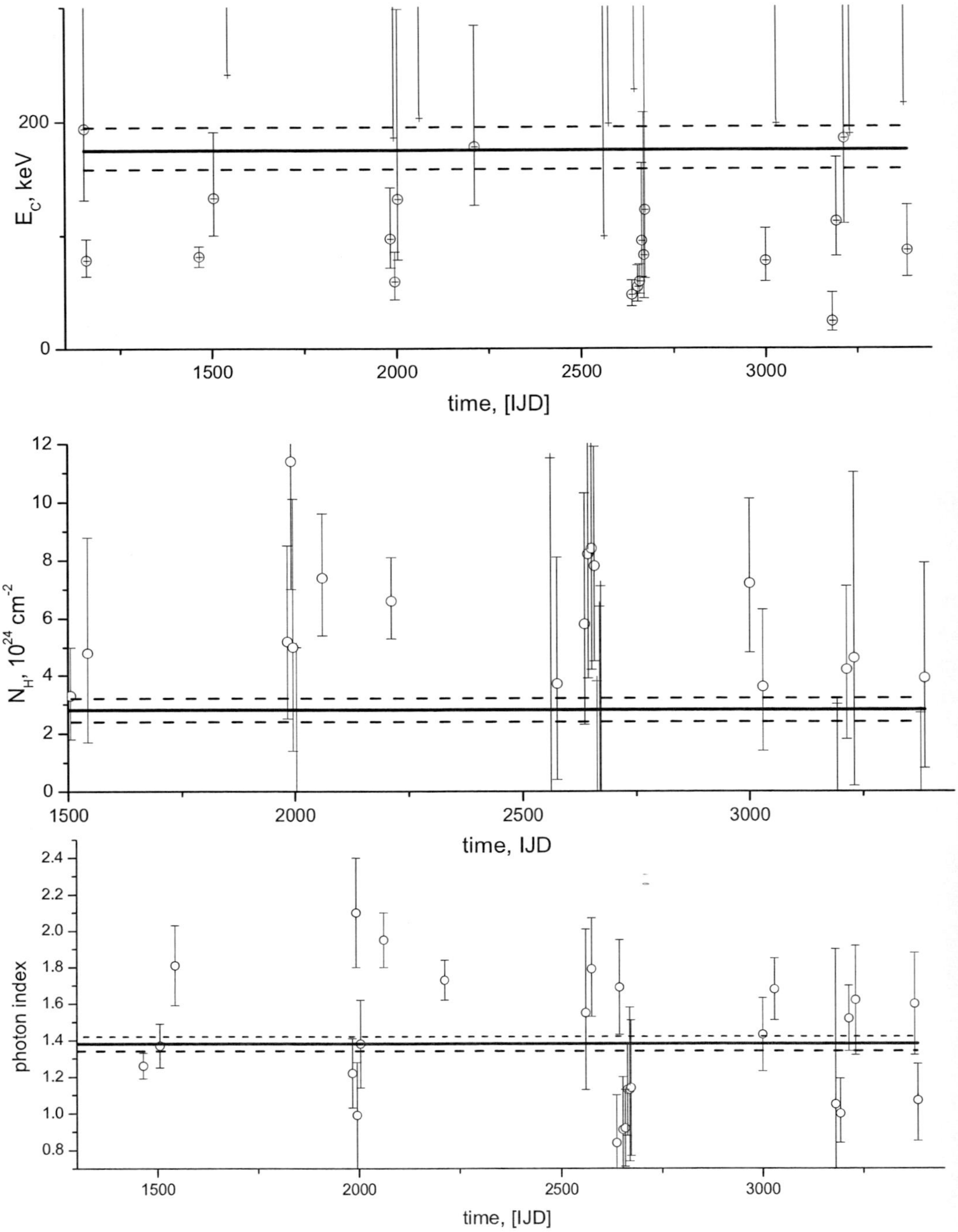

Figure 3. The best-fit parameters for partial data revolutions, depending from time.

Table 3. ***INTEGRAL*** **best-fit results for NGC 4945 during different phases. The errors are at the 90% confidence level.**

Phase, Parameter	OP	TP	"lo-cut"	"hi-cut"	"no-cut"
χ^2/d.o.f.	9.8/8	7.5/7	8.9/9	11.5/9	9.7/8
Γ	1.46 ± 0.04	1.17 ± 0.05	0.96 ± 0.05	1.25 ± 0.06	1.89 ± 0.11
E_C, keV	143^{+13}_{-12}	141^{+18}_{-15}	73 ± 5	140^{+22}_{-18}	> 580
N_H, $10^{24}\,\mathrm{cm}^{-2}$	5.0 ± 0.6	<1.0	3.4 ± 0.5	1.7 ± 0.8	$4.5^{+1.6}_{-1.5}$ [a]
$F^a_{20-40\,\mathrm{keV}}$	9.59 ± 0.07	$7.4^{+0.2}_{-0.1}$	9.3 ± 0.1	8.1 ± 0.1	7.2 ± 0.1
$F^a_{40-60\,\mathrm{keV}}$	7.8 ± 0.1	6.0 ± 0.1	5.8 ± 0.1	6.6 ± 0.1	5.8 ± 0.1
$F^a_{60-100\,\mathrm{keV}}$	9.8 ± 0.1	8.5 ± 0.1	8.0 ± 0.2	8.8 ± 0.1	8.0 ± 0.2
$F^a_{100-200\,\mathrm{keV}}$	10.4 ± 0.1	10.8 ± 0.2	11.8 ± 0.3	10.1 ± 0.2	11.8 ± 0.3
$F^a_{>200\,\mathrm{keV}}$	5.78 ± 0.05	7.7 ± 0.1	16.8 ± 0.4	6.1 ± 0.1	16.8 ± 0.4

flux given in units of $10^{-11}\,\mathrm{erg\,cm^{-2}\,s^{-1}}$

with the "lo-cut" and "high-cut" phases having harder spectra ($\Gamma = 0.98 \pm 0.05$ and $\Gamma = 1.25 \pm 0.06$) than the "no-cut" spectrum ($\Gamma = 1.89 \pm 0.11$).

Table 4. The correlation coefficients between the best-fit parameters and fluxes in several energy bands.

Parameter Energy band	Γ	E_C	N_H
20–40 keV	0.16	0.13	0.28
40–60 keV	0.39	0.21	0.09
60–100 keV	0.37	0.06	0.23
100–200 keV	-0.0007	-0.36	0.3
above 200 keV	-0.49	-0.68	0.18

We also had calculated the correlation coefficients between the three fitting parameters and fluxes in several energy bands. The correlation coefficients are shown in the Table 4. As we can see here, there is moderate anticorellation between the flux level in soft γ-rays (i.e. above 200 keV) and spectral shape (i.e. spectral slope and cut-off at high energies), but there is no prominent link between flux level absorbtion. This can be an another sign of a complex variability origins in 20–200 keV energy range.

4. Discussion

While our analysis covers time scales larger than a month, studies based on *Suzaku* investigated the variability on time scales shorter than a day [15, 16]. Their study revealed periodical variability of the primary continuum within hours. The variability observed with *Suzaku* was explained in terms of the reflection from cold neutral medium located at the distance from the AGN centre (i. e. probably the torus walls).

Comparably slow variations of the 20–200 keV flux and of the spectral slope are shown here based on *INTEGRAL* data. The flux level do not seem to be directly related to the spectral shape changes (except only the energies above 200 keV), as well as with absorbtion level.

4.1. Absorption

Changes of the column density of an obscuring medium are among the common causes for soft X-ray variability in Seyfert 2 [24]. In a Compton-thick AGN it can afflict the spectrum at higher energies, in our case, for instance, up to 40 keV. NGC 4945 is not an exception: absorption-induced variations were found in this object by Mariucci et al.[19]. The N_H changes founded here also confirm the *Suzaku* result that the absorbtion here is variable by factor of $\tilde{2}$ or greater.

The value of the column density in NGC 4945 obtained in our work is in the range $N_{\rm H} = (0.5 - 11.0) \times 10^{24}\,{\rm cm}^{-2}$, confirming that this source is a Compton thick type 2 AGN. Changes in the intrinsic absorption on the level mentioned above can explain the variability in the 10–40 keV band as shown by Mariucci et al. [19] studying *Suzaku* data. But the modest variability in absorption is not sufficient to explain the hard X-ray flux variations observed in the data presented here.

Also, the absorbtion varies independently from the spectral shape parameters (slope and cut-off), and the flux dependence on these parameters is quite complex. Thus we describe the spectral shape variability in the next subsection.

4.2. Spectral Slope and High-Energy Cut-Off

The best-fit model to the total exposure time spectrum is an absorbed power-law with the Compton reflection component ($R = 0.7$), a high-energy cut-off near 175 keV and a spectral slope of $\Gamma = 1.38$.

However, as we can see from the results of our spectral modelling obtained during the independent modelling of different data revolutions, the shape of the 20–500 keV spectrum of NGC 4945 varies both in photon index and cut-off energy, as well as in the absorbtion level. The lowest value of cut-off energy obtained here is $E_C \simeq 25^{+25}_{-9}$ keV.

In NGC 4945 we observe at least three different types of variations:

- 10–40 keV variability due to an obscuring medium density changes [19];
- fast intraday variations of the hard X-ray emission above 10 keV, likely due to the reflection changes in the obscuring torus [16];
- variations of the spectral shape (i.e. both the photon index and the high energy cut-off) in the 20–500 keV energy range (this work).

The possible origins of the latter are discussed in the next subsection, when comparing it with other objects with similar variability.

4.3. NGC 4945 Among Others vs. Spin-Paradigm

An example of variations like the one we observe in NGC 4945 can be found in other two radio-quiet Seyfert 1 galaxies IC 4329A [28] and NGC 4388 [9]. In the first object a high-energy cut-off near 40 keV was found in the August 2003 *INTEGRAL*/ISGRI observations but no cut-off was detected in July 2003. At the same time, the flux did not show significant variability [28] . In the second the phases with HEC below 100 keV were disclosed as well as the phases with no HEC. The spectral shape in NGC 4388 is also not correlated with a flux level [9].

A cut-off below 100 keV was seen also in NGC 5506 [26]. NGC 5506 has cut-off values in the range 40 – 100 keV and a softer ($\Gamma = 1.9$) average spectrum than NGC 4945. All the three objets mentioned above are the RQs. NGC 4945 is also an RQ one [8]. However here we can observe the non-stable spectral shape with with temporal features prescribed by spin-paradigm rather to RL objects.

For AGN, as well as for Galactic Black Hole systems, the hard spectrum and high-energy cut-off is supposed to be connected with radio-loudness and jets [14]. NGC 5506, NGC 4388, IC 4329A and NGC 4945 are RQ AGNs [10].

This brings up the question about the radio emission of this AGN - is it permanently weak or are there periods of comparatively high radio luminosity? Is the radio luminosity of this object connected to the flux and the spectral parameters in the X-ray band?

5. Conclusion

We have performed an analysis of the *INTEGRAL* data obtained for NGC 4945 during the period March 2003 to August 2009. The main results based on the data obtained during the six years of *INTEGRAL* observations of NGC 4945 can be summarized as follows:

- the uncorrelated flux variability in the 20–40 keV, 40–60 keV and 60–100 keV energy bands on timescales of months;
- significant spectral shape changes of the 20–500 keV spectrum;
- the lower value of the exponential cut-off at high energies (found during the hard spectrum phase) is similar to that of radio loud AGN despite NGC 4945 is an RQ object.

The spectral shape variability is not connected with changes of the intrinsic absorption. In GBH systems the hard/soft variations are supposed to be connected with radio ejections/jet activity. Thus it would be interesting to test this link in AGN, e.g. to observe NGC 4945 simultaneously in hard X-rays/γ-rays and at radio frequencies to trace out the link between the radio luminosity and spectral shape in X-rays/γ-rays.

References

[1] R.Antonucci, F. P. Miller *ApJ*,1985, Vol.297, P621-632.

[2] R. Antonucci, J.S. Ulvestad *ApJ*, 1985, Vol. 294, P.158-182.

[3] K. A. Arnaud In: Astronomical Data Analysis Software and Systems V, ASP Conf. Series, eds. G. Jacoby and J. Barnes, 1996, Vol.101, P.17.

[4] R. D. Blandford In: Astrophysical Disks, ASP Conf. Series, ed. J. A. Sellwood and J. J. Goodman, 1999, Vol. 160, P265.

[5] R. Blandford, D.G. Payne *MNRAS*, 1982, Vol.199, P.883-903.

[6] R. Blandford, R. Znajek *MNRAS*, 1977, vol. 179, P.433-456.

[7] T.J.-L.Courvoisier, R. Walter, V. Beckmann, et al. *A&A*, 2003, Vol.411, P.L53-L57.

[8] J.S. Dunlop 2001 In: QSO hosts and their environments, Proceedings of an international workshop, Dordrecht: Kluwer Academic / Plenum Publishers, Ed. by I.Mrquez, J.Masegosa, A.Del Olmo, L. Lara, E. Garca, J. Molina, 2001, Vol.xvii, P.3.

[9] E. Fedorova, V. Beckmann, A. Neronov,S. Soldi *MNRAS*, 2011, Vol. 417, Is.2, P.1140-1147.

[10] J.F. Gallimore, D.J. Axon, C.P. o'Dea, S.A. Baum, A. Pedlar AJ, 2006, Vol.132, P.546-569.

[11] D. Garofalo,D.A. Evans, R.M. Sambruna *MNRAS*, 2010, Vol.406, is.2, P.975-986.

[12] M. Guainazzi, G. Matt, W.N. Brandt et al. *A&A*, 2000, Vol.356, P463-474.

[13] J. I. Harnett, R. Wielebinski, R. F. Haynes, U. Klein *A&A.*, 1989, Vol. 216, N1-2, P39-43.

[14] L. C. Ho *Ap&SS*, 2005, Vol.300, is.1-3, P.219-225.

[15] N. Isobe, A. Kubota, K. Makishima, P. Ganghi et al. *PASJ*, 2008, Vol.60, NoSp1, P.S241-S250.

[16] T. Itoh, C. Done, K. Makishima, G. Madejski et al. *PASJ*, 2007, Vol.60, NoSP1, P.S251-262.

[17] J.P. Lenain, C. Ricci, M. Tuerler, D. Dorner, R. Walter *A&A*, 2010, Vol.524, id.A72.

[18] P. Magdziarz, A.A. Zdziarski *MNRAS*, 1995, Vol.273, P837-848.

[19] A.Marinucci, G. Risali, S. Bianchi, M. Elvis et al. The unique Suzaku discovery of variability in Compton-thick absorber in NGC 4945, Suzaku 2011, *SLAC*, 20-22 July 2011.

[20] A. Merloni, A. C. Fabian *MNRAS*, 2002, Vol.332, Is.1, P.165-175.

[21] R. McLure, J. Dunlop In: QSO hosts and their environments, Proceedings of an international workshop, Dordrecht: Kluwer Academic / Plenum Publishers, Ed. by I. Mrquez, J. Masegosa, A. Del Olmo, L. Lara, E. Garca, J. Molina, 2001, Vol.xvii, P.27.

[22] P. Padovani, C.M. Urry *ApJ*, 1991, Vol.368, P.373-379.

[23] G. Risaliti *A&A*, 2002, Vol.386, P.379-398.

[24] G. Risaliti 2009, Proc. IAU Symp., Vol.267, P.119.

[25] N. Schurch,T. Roberts, R. Warwick, *MNRAS*, 2002, Vol.335, is.2, P.241-246.

[26] S. Soldi, V. Beckmann, N. Gehrels,S. De Jong, P. Lubiński In: proc. of the Workshop "Narrow-Line Seyfert 1 Galaxies and Their Place in the Universe", 2011, arXiv:1105.5993.

[27] S. Soldi, V. Beckmann, L. Bassani,T.J.-L. Courvoisier et al. *A&A*, 2005, Vol.444, is.2, P.431-441.

[28] S. Soldi,G. Ponti, V. Beckmann, P. Lubiński 2010, in: "The Extreme Sky: Sampling the Universe above 10 keV",

[29] J. B. Whiteoak, W. E. Wilson *MNRAS*, 1990, Vol.245, P.665.arXiv:1001.4348.

[30] A.S. Wilson, E.J.M. Colbert, J. Astroph 1995, Vol.438, P62-71.

[31] J.-H. Woo, C.M. Urry *ApJ*, 2002, Vol.581, P.L5-L8.

In: Galaxies: Evolution, Morphology and Dynamics
Editors: A. Guerrera Sanchez et al., pp. 149-163
ISBN: 978-1-62081-185-6

Chapter 6

LINDBLAD'S EPICYCLES — VALID METHOD OR BAD SCIENCE?

*Charles Francis**
Hastings, UK

Abstract

The study of Galactic orbits in the last eighty years has been dominated by statistical assumptions made because of the lack of empirical evidence available in the early 20th century. Using evidence from Hipparcos and recent radial velocity surveys, Francis and Anderson recently showed that spiral structure is primarily a consequence of gravitational alignments of stellar orbits. I review the mechanism which creates spiral structure, consider the validity of widely held assumptions in galactic dynamics and the implications to notions such as the asymmetric drift and disc heating. I identify a number of fundamental mathematical and physical errors in Lindblad's epicycle theory and in density wave theory. Students should be made aware that these ideas can no longer be considered as science, and authors of textbooks should consider whether they merit anything more than a historical note.

PACS 95.10.Ce; 98.10.+z; 98.35.Df

Keywords: Astrometry – celestial mechanics – stars: kinematics – Galaxy: kinematics and dynamics – Galaxy: solar neighbourhood.

1. Background

In popular culture epicycles have become almost synonymous with bad science; "adding epicycles" refers to a process of introducing fudges to make a theory fit data, when actually the theory needs to be replaced in its entirety. It is generally believed that epicycles were banished from science when Newton solved his equations of motion and showed that it follows from the inverse square law of gravity that planetary orbits are ellipses. So, it comes as something of a surprise to those unfamiliar with galactic dynamics that the galactic orbits of stars are treated in textbooks using a theory of epicycles revitalized by Bertil Lindblad in

*E-mail address: charles.e.h.francis@googlemail.com

the 1920s, and used as the basis of density wave theory, which, as reinforced by Lin, Yuan and Shu (1969) and by Kalnajs (1973), has been the leading model of spiral structure for nearly 40 years. Of course, it is perfectly possible to approximate any orbit with a system of epicycles; the question is whether doing so has either quantitative or qualitative benefit.

Galactic orbits are expected to precess because of the matter distribution in the disc and the halo, generating a rosette. Many studies in stellar dynamics rest upon the notion that stars move in near-circular orbits and that their orbits precess at different rates, resulting in a well mixed distribution. E.g. the calculation of the LSR from Strömberg's asymmetric drift equation (e.g. Dehnen & Binney, 1998). However, there have been a number of recent studies challenging the assumption of a well mixed distribution (Skuljan, Hearnshaw & Cottrell, 1999; Fux 2001; Dehnen, 1998; Famaey et al., 2005, Chakrabarty, 2004 & 2007, Quillen 2003, Minchev & Quillen 2006, de Simone et al, 2004, Chakrabarty & Sideris, 2008). In a recent study, Francis and Anderson (2009b; herein FA09b) described how mutual gravity between stars causes orbits to align on approximately logarithmic spirals, and showed from observations of the galactic gas distribution and the motions of over 20 000 stars in the solar neighbourhood that the Milky Way is a two-armed spiral. Contrary to conventional wisdom, stars do not move through the arms on near-circular orbits, but move along an arm on the inward part of their orbits, leave the arm soon after pericentre, cross the other arm on the outward part and rejoining the original arm before apocentre (a brief review is given in section 2). The distribution does not become well mixed over the course of time. Instead orbital rosettes become aligned with spiral arms. As a result, a number of conventional calculations fail (section 2.8).

Epicycles have only been used to study orbits in galactic dynamics. In other fields, perturbations to an orbit due to alteration from the potential due to a central mass are studied using the eccentricity vector or the Laplace-Runge-Lenz vector (see e.g. Arnold, 1989; Goldstein, 1980). It is appropriate, therefore, to assess whether Lindblad's epicycles have any value in the study of galactic dynamics. Section 3 will describe Lindblad's epicycles, and show that the complexity of the treatment has no benefit over a more conventional analysis, but has rather obscured dynamics and lead to mistakes in analysis.

2. Gravitationally Aligned Rosettes

2.1. The Eccentricity Vector

For an elliptical orbit the eccentricity vector is defined as the vector pointing toward pericentre and with magnitude equal to the orbit's scalar eccentricity. It is given by

$$\boldsymbol{e} = \frac{|\boldsymbol{v}|^2\boldsymbol{r}}{\mu} - \frac{(\boldsymbol{r}.\boldsymbol{v})\boldsymbol{v}}{\mu} - \frac{\boldsymbol{r}}{|\boldsymbol{r}|}, \tag{1}$$

where $\boldsymbol{v}$ is the velocity vector, $\boldsymbol{r}$ is the radial vector, and $\mu = GM$ is the standard gravitational parameter for orbits about a mass, M. For a Keplerian orbit the eccentricity vector is a constant of the motion. Stellar orbits are not strictly elliptical, but the orbit will approximate an ellipse at each part of its motion, and the eccentricity vector remains a useful measure (the Laplace-Runge-Lenz vector, which is the same up to a multiplicative factor, is also used to describe perturbations to elliptical orbits).

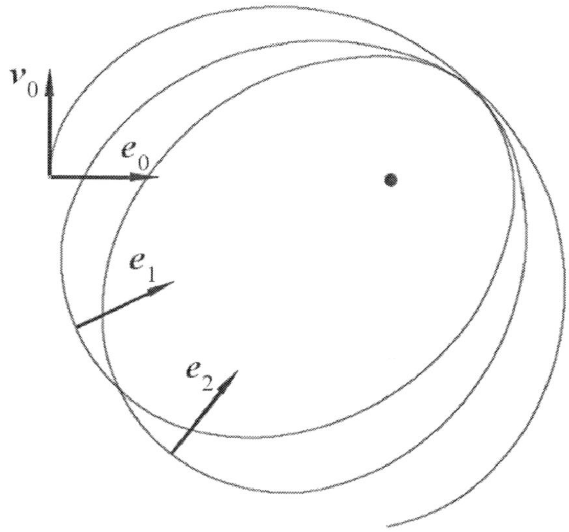

Figure 1. The eccentricity vector of an orbit regresses for a central core plus disc. Regression has been exaggerated by increasing the mass of the disc relative to the core (by comparison with the Milky Way). The simulation used a central mass of 35 billion solar masses, a disc density $0.3e^{-R/3}$ billion solar masses per kpc^2, initial radius 8kpc and initial velocity 190 km s^{-1}.

2.2. Orbital Precession

In the solar system, mass is concentrated at the Sun, and planetary orbits are ellipses. In a spiral galaxy mass is distributed throughout the disc and in the halo. As a star moves towards pericentre, the gravitational mass drawing it towards the galactic centre is less than it would be if all the mass of the galaxy was concentrated at one central point. As a result, the orbit of the star is less curved at pericentre than an ellipse, and the orbit precesses, forming a rosette. We used a numerical simulation to model an orbit for a galaxy with a central core plus a disc with an exponentially decreasing surface density (figure 1). Imagine looking at the motion from above, from a platform rotating at the rate of precession; the orbit will be approximately elliptical with the galactic centre at the focus.

2.3. Spiral Structure

An equiangular spiral structure can be constructed by enlarging an ellipse by a constant factor, k, centred at the focus and rotating it by a constant angle, τ, with each enlargement (figure 2). The pitch angle of the spiral depends only on k and τ, not on the eccentricity of the ellipse. The pitch angle of a given spiral galaxy is directly related to the orbital eccentricities of stars in that galaxy. For a given pitch angle, ellipses with a range of eccentricities can be fitted to the spiral, depending on how narrow one wants to make the spiral structure and what proportion of the circumference of the ellipse one wants to lie within it. Higher eccentricity orbits fit spirals with higher pitch angles. Thus we can understand

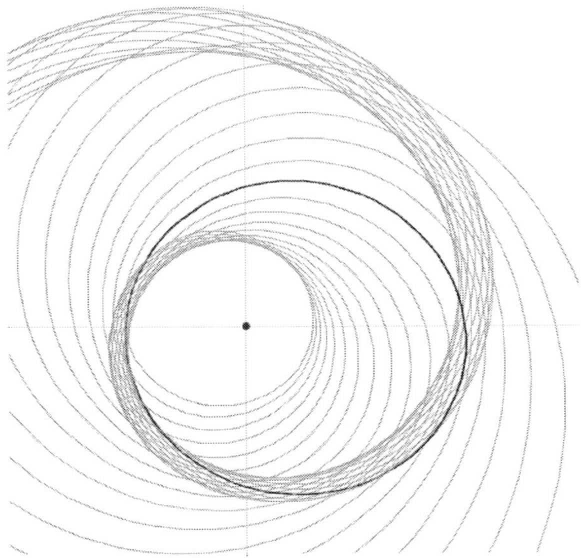

Figure 2. An equiangular spiral with a pitch angle of 11, constructed by repeatedly enlarging an ellipse with eccentricity 0.3 by a factor 1.05 and rotating it through 15 with each enlargement. Lower eccentricity ellipses produce a narrower structure. Ellipses with eccentricity greater than about 0.25 have more than half their circumference within the spiral region. Ellipses with eccentricity greater than about 0.35 produce probably too broad a structure to model a spiral arm with this pitch angle, but give a good fit for spirals with greater pitch angle.

the spiral structure of the arms, provided that a mechanism can be exhibited which causes orbits to align with the arm, and which ensures that the angular rate of orbital precession is independent of orbital radius.

2.4. Spiral Potential

The gravitational potential of a spiral galaxy can be described as a giant, spiral-grooved funnel (figure 3). The grooves represent the gravitational field of the spiral arms. A star near apocentre, the slowest part of its orbit, will tend to fall into a groove and then follow the groove, picking up momentum as it goes. Eventually, the star gains enough momentum to jump free of its groove. It crosses over the next-highest groove, then falls back to a higher point in its original groove (an animation is shown at http://rqgravity.net/SpiralStructure). At the same time, the funnel rotates slowly backwards due to orbital precession.

As stars are drawn into an arm, the gravitational field of the arm grows stronger, drawing greater number of stars into the arm. This mutual gravity between stars reinforces spiral structure, and the potential field of the arms locks the rate of orbital precession to spiral pattern speed for a wide range of orbits. FA09b describes the evolution from flocculent through multiarmed to grand-design bisymmetric spirals.

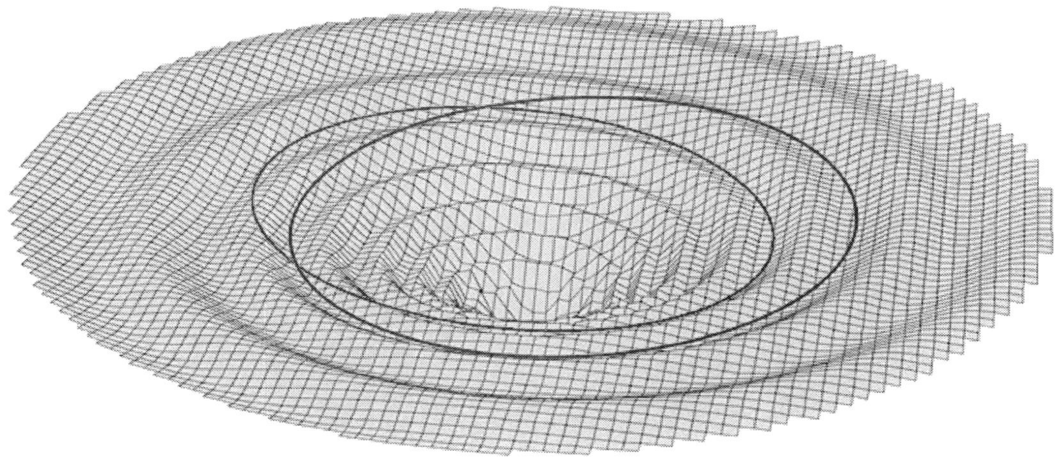

Figure 3. The gravitational potential of a bisymmetric spiral galaxy, showing the alignment of elliptical orbits with troughs in the potential.

2.5. Star Formation in Spiral Arms

Under gravity, gas clouds follow similar orbits to stars (figure 4). Gas in the arm is in turbulent motion, as gas clouds seek to cross in the arm and gain velocity as they approach pericentre. Whereas stars rarely collide because of their small size compared to space between them, when outgoing gas from one arm meets ingoing gas in another arm, collisions between gas clouds create regions of higher pressure, and greater turbulence. Pockets of extreme pressure due to turbulence generate the molecular clouds in which new stars form.

2.6. Formation of Bisymmetric Spirals

In a multi-arm spiral, outgoing gas meeting an arm would have greater mass than ingoing gas in the arm. This would tend to remove gas from the arm. In a two armed spiral, the gas in the arm has greater mass. Thus, a two-armed gaseous spiral can be stable, whereas multi-armed gaseous spirals cannot. Outgoing gas applies pressure to the trailing edge of a spiral arm. If one gaseous arm advances compared to the bisymmetric position, the pressure due to gas from the other arm will be reduced. At the same time, pressure on the retarded arm due to outgoing gas from the advanced arm will be increased. Thus gas motions preserve the symmetry of two-armed spirals.

2.7. Spiral Structure of the Milky Way

Pitch angles of around 10-15 (e.g. figure 2), as used in four armed spirals (e.g. Georgelin & Georgelin, 1976; Russeil, 2003; Levine, Blitz and Heiles), correspond to orbital eccentricities in a range greater than about 0.25, and are incompatible with the eccentricity distribution of the local stellar population (the eccentricity distribution in figure 5 is based on the LSR calculated by Francis and Anderson, 2009a, but the incompatibility remains for any reasonable value of the LSR). A two-armed spiral with pitch angle around 5 gives a

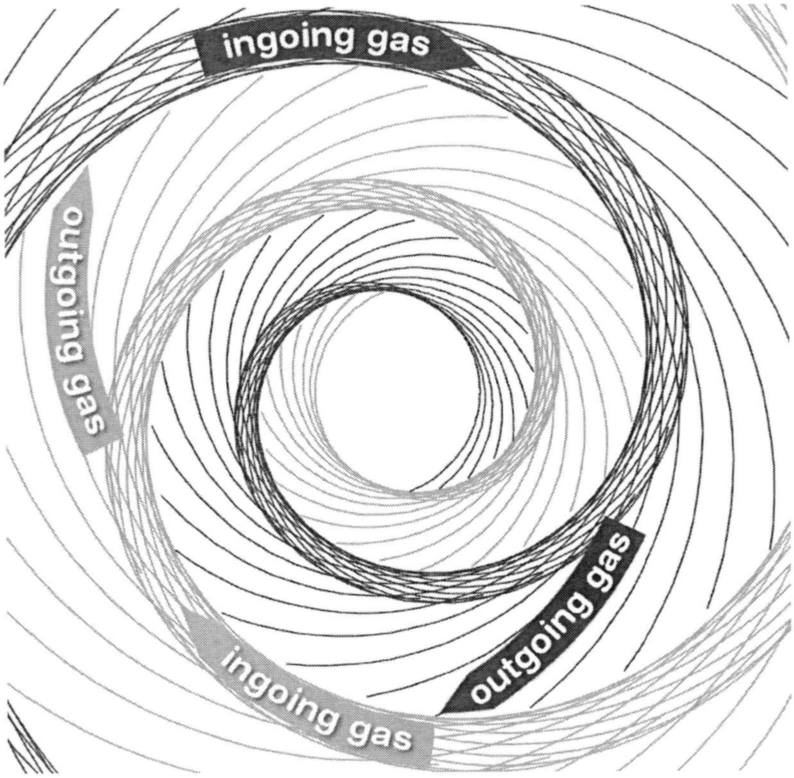

Figure 4. Gas motions in a bisymmetric spiral galaxy.

good fit to the bulk of observed eccentricities in a range from about 0.1 to about 0.2, and also gives a good fit with the observed HI distribution (FA09b).

A two-armed spiral necessitates a little care to avoid confusion in naming the arms, because traditionally named sectors with the same name lie on different arms (figure 6). Orion is not a separate spur, but is a part of a major arm connecting Perseus in the direction of rotation to Sagittarius in the direction of anti-rotation. We have called this major spiral arm the Orion Arm. The Orion arm contains Norma, Perseus, Orion, Sagittarius, and Cygnus sectors. The Centaurus arm contains Sagittarius, Scutum-Crux, Cygnus, and Perseus sectors. The solar orbit is shown in approximation, together with its major axis and latus rectum.

2.8. The Structure of the Local Velocity Distribution

The velocity distribution (figure 7) contains motions of new stars (Pleiades stream), a flow of stars in the spiral arm on the inward part of their orbits, stars on the outward part of their orbit crossing the Orion arm (Hyades stream), moderately young stars which have not achieved typical orbits in the arm (Sirius stream) and older stars in more eccentric orbits, (Hercules and Alpha Ceti stream). Thus the distribution is highly structured, and traditional

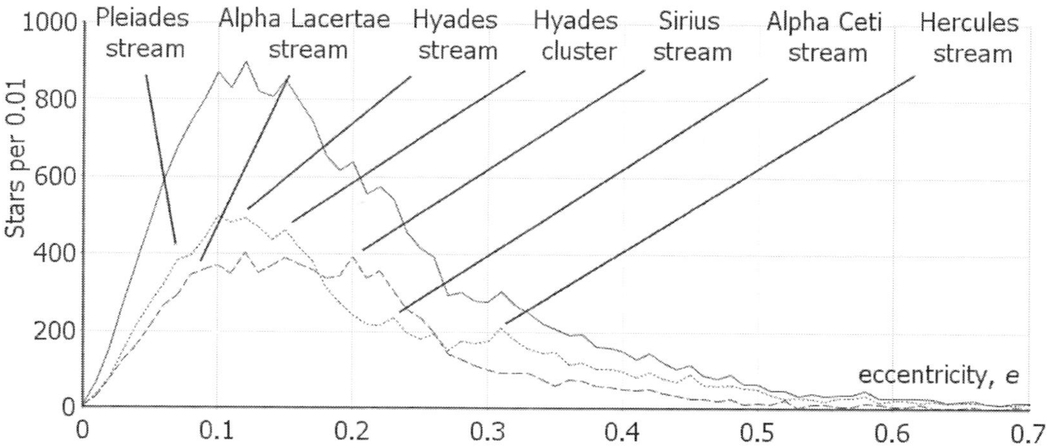

Figure 5. Eccentricity distribution (from FA09a) for stars closer to apocentre (dotted) and stars closer to pericentre (dashed), as defined by position with respect to the semi-latus rectum.

measures such as dispersion and mean velocity have little meaning for the population as a whole. As a result, it is necessary to put aside typical text book analyses.

For example, the Oort constants have no meaning, the LSR cannot be calculated by estimating the asymmetric drift, and there is no evidence for disc heating — the hypothetical process by which scattering events are supposed to cause the random velocities of stars to increase with age (Jenkins, 1992). There is an increase in velocity dispersion with colour, up to Parenago's discontinuity (Parenago, 1950). Dehnen & Binney (1998) suggest that the reason for Parenago's discontinuity is the heating of the disc. However, disc heating appears to contradict the second law of thermodynamics and relies on the mistaken belief that velocity dispersion is a measure of randomness. The truth is that dispersion rises with age because old stars can be found in stable orbits with greater than normal eccentricity, while young stars rarely enter these orbits (FA09b).

3. Errors in Density Wave Theory

3.1. The Winding Problem

The analysis of spiral structure is often introduced with a description of the winding problem. Lindblad argued that the arms cannot consist of stars orbiting together in approximately circular orbits, because orbital velocity decreases with orbital radius, so that a spiral consisting of regions of greater density would wind up over a period of time. Unfortunately, by considering the orbital velocity of stars, this analysis has misdirected later investigations. Stars are not bound in any form of solid structure. Nor is there any prior reason to assume that their orbits can be treated as nearly circular. The evidence of near-circular orbits for moons and planets in the solar systems suggests that tidal forces lead to a decrease in eccentricity over large numbers of orbits, but typical stars at the solar radius will have orbited

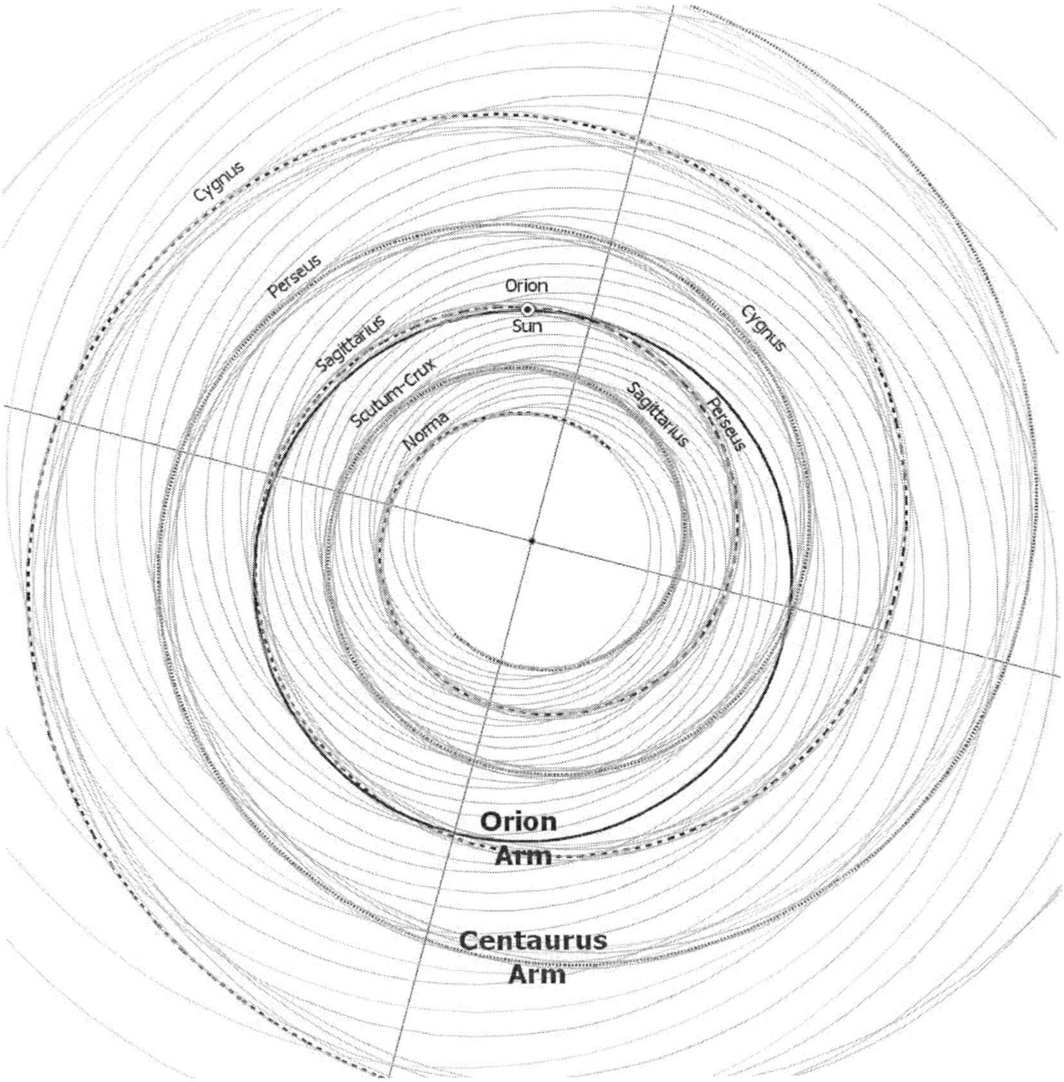

Figure 6. Two-armed spiral model of the Milky Way, based an angular increment of t = 30 for each 105% enlargement, giving a pitch angle of 5.44. The solar eccentricity, 0.138, has been used for the diagram.

the Milky way less than about 100 times — far too few to expect the system to be close to a circular equilibrium.

Spiral structure as described in section 2 is in fact independent of orbital velocity; the winding problem as described by Lindblad does not apply. Stars do not individually come together and share similar orbits as would be suggested by a solid motion. Rather stellar paths align on the spiral arm for much of the orbit. The behaviour of the spiral over time is then determined by the rate of precession of the orbits, not by orbital velocity. The gravitational potential of the spiral modifies the precession of the orbits, so as to preserve and reinforce spiral structure, and to ensure that they all regress at the same rate, the rate of

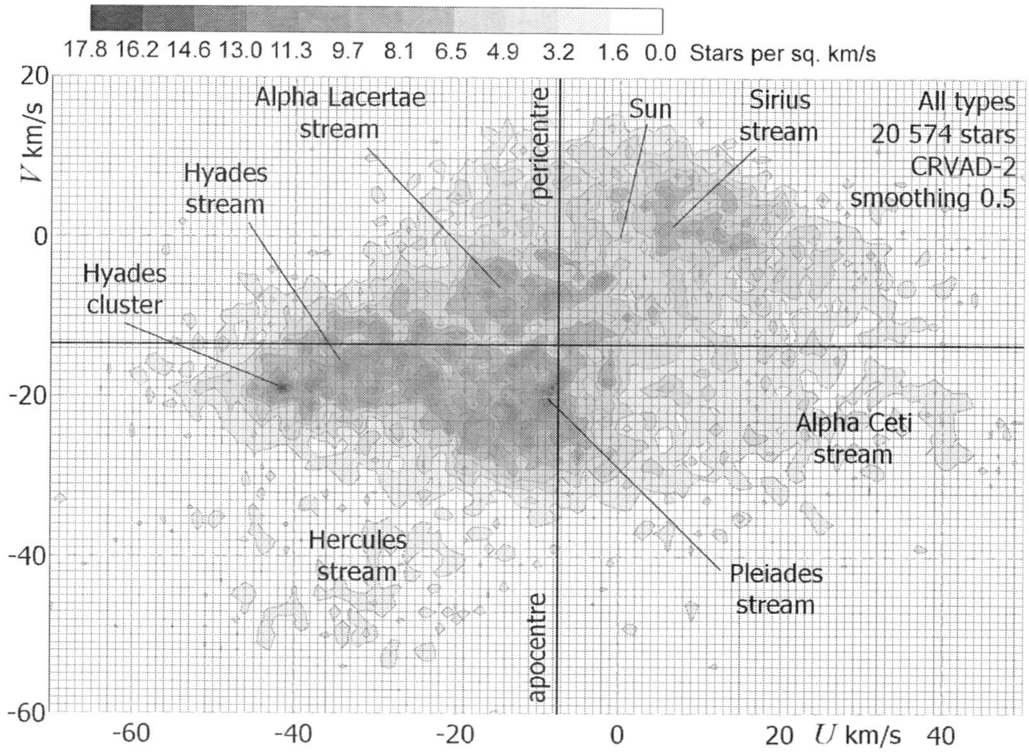

Figure 7. The velocity distribution in the Galactic plane from FA09a, showing the Hyades, Pleiades, Sirius, Alpha Ceti, Alpha Lacertae and Hercules streams.

spiral pattern speed.

3.2. The Epicyclic Approximation

Lindblad's epicycles perturb a circular orbit by superimposing an elliptical motion. As is seen in figure 8, for low eccentricities it is possible to approximate a precessing orbit. However, this is not useful. After the effort, and with the inaccuracy, of introducing the epicyclic approximation, no more has been said than that the orbit is a rosette. To take the analysis any further, and calculate, for example, the rate of precession, would require further corrections because the epicyclic approximation is only good for low eccentricities, and would require knowledge of the gravitational potential throughout the orbit. Epicycles cannot be used as an iterative method, because there is no practicable way to continue the iteration. If fluctuations in potential due to spiral arms, the bar, and satellites, are to be considered, then analysis is only possible by numerical methods. However, if a computer will be used to provide a numeric solution, there is no point in starting with an approximation. It would be better, and simpler, to find a numerical solution directly from the equations of Newtonian gravity.

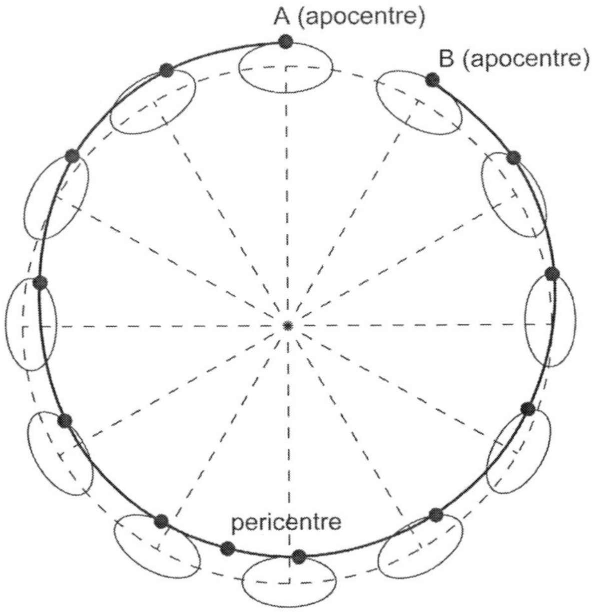

Figure 8. Lindblad's epicycles approximate a precessing ellipse by superimposing an elliptical motion on a circular orbit.

3.3. Closed Orbits

A closed orbit is one which returns to apocentre at the same point after a number of cycles, and then repeats the same path. We can close the orbit in figure 4 by using coordinates rotating at just such a rate that apocentre moves so as to coincide a previous apocentre after a number of orbits. We can calculate an equation for the rate of rotation of the coordinate system as follows:

Angular speed, or angular frequency, of circular motion: Ω.
Angular speed of radial oscillation (i.e. elliptical motion): κ.
Period of radial oscillation: $1\pi/\kappa$.
Angular distance, AB, after $m > 0$ radial oscillations:

$$\mathsf{AB} = 2\pi n \pm \Omega \frac{2\pi m}{\kappa}, \quad n \in \mathbb{Z}. \tag{2}$$

Mistake 1: Standard treatments (e.g. Binney & Tremaine, 1987; Binney & Merrifield 1998; Carroll & Ostlie, 1996) overlook the possibility of a minus sign in eq. (2). The orbit may be closed by rotating through either the major or the minor arc AB.

Angular speed of rotating reference frame: f.

To establish a closed orbit in the rotating coordinates, after m periods of the radial oscillation, we require,

$$f\frac{2\pi m}{\kappa} = \mathsf{AB} = 2\pi n \pm \Omega\frac{2\pi m}{\kappa}, \tag{3}$$

$$f = \frac{n}{m}\kappa \pm \Omega. \tag{4}$$

The usual solution uses plus and $m = -2n$.

$$f = \Omega - \frac{1}{2}\kappa. \tag{5}$$

Mistake 2: This ignores imaging issues with $m > 1$, such that the rotating coordinate system shows more than one orbit.

Mistake 3: The use of plus in eq. (4) means the rotation of coordinates is in the direction of orbital motion, but actually orbits regress.

The correct rate for spiral pattern speed requires minus and $m = n = 1$. Hence

$$f = \kappa - \Omega. \tag{6}$$

3.4. Density Wave Theory

Spiral structure is usually explained using the density wave theory of Lin et al. (1969), according to which stars move through the arms on near-circular orbits, and the arms consist of dense regions analogous to regions of heavy traffic on a motorway.

Mistake 4: A simple analogy with patches of heavy traffic fails because a wave effect would require that stars slow down when they approach a dense region, but the gravity of the dense region would cause them to speed up. Even if the theory were right, the analogy would be wrong.

Mistake 5: The claim that stars move through the arms on near-circular orbits is empirically incorrect (Francis & Anderson, 2009a)

Following Kalnajs (1973), density wave theory is usually explained by means of a diagram (figure 5) constructed by enlarging and rotating ovals. The orbits are an epicyclic approximation in coordinates rotating at a rate $\Omega - \kappa/2$ and appear as ovals centred at the galactic centre.

Mistake 6 (following from mistake 2): Even if the rotation of the ovals were correct (it is not), because of the rate of rotation of coordinates in which it is drawn, figure 5 would show a single spiral twice, not a bisymmetric spiral.

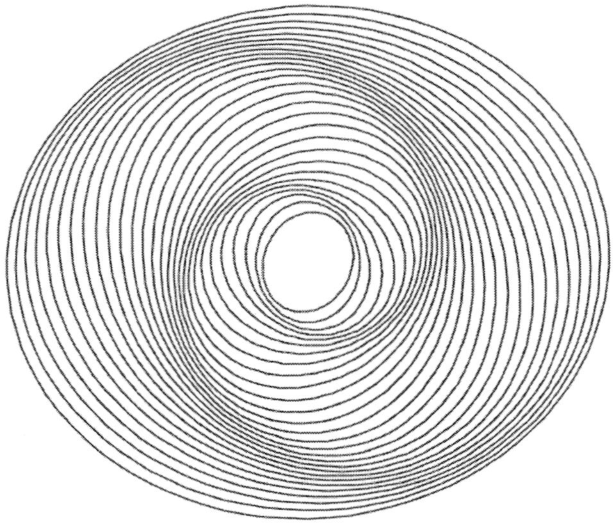

Figure 9. Enlarging and rotating ovals aligned at the centre generates a two-armed spiral.

Mistake 7 (to be replaced by mistake 8): The increase in density shown as spirals in the figure represents only a small proportion of the orbit. If stars were placed randomly, one on each oval, this would lead to a very small increase in stellar density on the arms, an order of magnitude less than is observed.

Mistake 8: (correcting mistake 7): The orbits in the figure are imagined as gas in lamina flow. Then the increase in gas density in the spiral pattern is assumed to instigate star creation in the arms. But this increase in density would not be sufficient to initiate star formation, and spiral arms contain a substantial increase in density of old as well as new stars.

Mistake 9: (following from mistakes 6 & 8): If the ovals were to represent gas motion, it could not be treated as lamina flow, because while paths do not cross in rotating coordinates, they do cross in physical space (figure 6). Taken as the motions of gases, figure 5 is impossible.

4. Conclusion

Understanding the cause of spiral structure has appeared a difficult problem for about eighty years, ever since it was clearly recognised that spiral nebulae are other galaxies. It is a many-body problem with unknown initial conditions, and its solution involves the turbulent motion of interstellar gas in addition to the motions of the stars. Nonetheless, once known, the solution is remarkably straightforward, and is confirmed by the empirical evidence of the local velocity distribution and the neutral hydrogen distribution.

The introduction of a working model for spiral galaxies will lead to important changes in the study of galactic dynamics. Many of the treatments found in textbooks are seen not

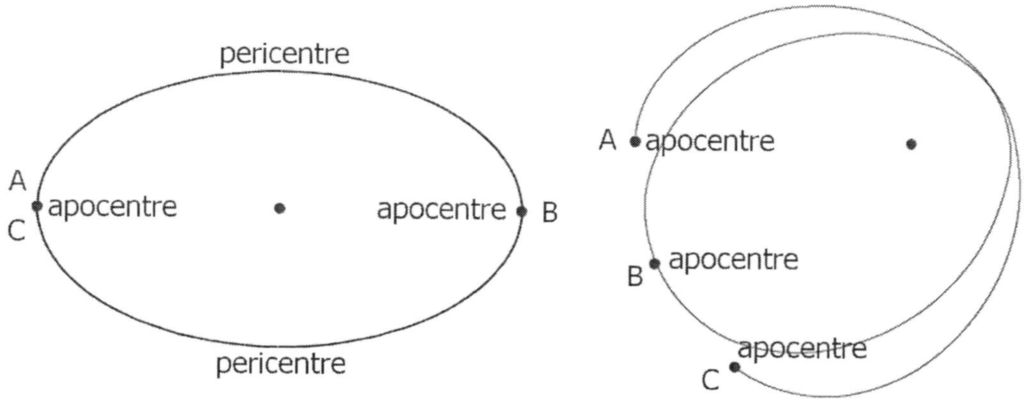

Figure 10. (left) Closed orbit in rotating coordinates with $f = \Omega - \kappa/2$, and (right) the corresponding path in non-rotating coordinates.

to apply. Perhaps of greater concern to astrophysics as a science is that some of those treatments are not even internally consistent. When mathematicians have addressed the problem of orbits in a gravitational field differing from that of a central mass distribution, they have perturbed the eccentricity vector or the Laplace-Runge-Lenz vector. Lindblad's epicycles have distracted astrophysicists from this type of analysis and substituted a model which is conceptually more complex, and which, because numerical solution is still required, adds nothing to what is already known of galactic orbits. The analysis has been based empirical assumptions about orbits which were not justified from data at the time, and which have since proven false, and it contains elementary mathematical mistakes which have been compounded by further mistakes in the development of density wave theory.

The implication to astrophysics is severe. The motions of stars are governed by known mathematical laws. Astrophysics is, or at least it should be, a mathematical science. One should therefore expect that theories in astrophysics are subjected to rigorous mathematical scrutiny. Regrettably, the degree of scrutiny applied to Lindblad's epicycles and to density wave theory has been seriously lacking. Students should be made aware that these ideas can no longer be considered as science, and authors of textbooks should consider whether they merit anything more than a historical note.

References

V.I. Arnold, 1989, *Mathematical Methods of Classical Mechanics*, 2nd ed., New York: Springer-Verlag. p. 38.

J.J. Binney & M. Merrifield, 1998, *Galactic Astronomy,* Princeton University Press.

J. Binney & S. Tremaine, 1987, *Galactic Dynamics,* Princeton University Press.

B.W. Carroll & D.A. Ostlie, 1996, *Modern Astrophysics,* International Edition, Addison-Wesley.

D. Chakrabarty, 2004, *MNRAS*, **352**, 427.

D. Chakrabarty, 2007, *A.&A.*, **467**, 145.

D. Chakrabarty and I.V. Sideris, 2008 *A.&A.*, **488**, 161.

W. Dehnen, 1998, *Ap. J.*, **115**, 2384.

W. Dehnen & J.J.Binney, 1998, *MNRAS* **298**, 387.

B. Famaey, A. Jorissen, X. Luri, M. Mayor, S. Udry, H. Dejonghe and C. Turon, 2005, Local Kinematics of K and M Giants from CORAVEL/Hipparcos/Tycho-2 Data, *A.&A.*, **430**, 165.

C. Francis & E. Anderson, 2009a, Calculation of the Local Standard of Rest from 20,574 Local Stars in the New Hipparcos Reduction with known Radial Velocities, *New Astronomy*, **14**, pp. 615-629.

C. Francis & E. Anderson, 2009b, Galactic spiral structure, *Proc. R. Soc. A,* **465**, 3401-3423, doi: 10.1098/rspa.2009.0036.

R. Fux, 2001, *A.&A.*, **373**, 511F.

Y.M. Georgelin & Y.P. Georgelin, 1976, The spiral structure of our Galaxy determined from H II regions, *A.& A.*, **49**, 1, p.57-79.

H. Goldstein, 1980, *Classical Mechanics* (2nd ed.). Addison Wesley. pp. 102-105, 421-422.

A. Jenkins, 1992, *MNRAS*, **257**, 620.

A.J. Kalnajs, 1973, Spiral structure viewed as a density wave, *Proc. Astr. Soc. Aust.*, **2**, p174 -177.

C.C. Lin, C. Yuan and F.H. Shu, 1969, On the Spiral Structure of Disk Galaxies III. Comparison with Observations, *Ap.J.* **155**, 721-746.

E.S. Levine, L. Blitz and C. Heiles, 2006, The Spiral Structure of the Outer Milky Way in Hydrogen, *Science*, **312**, 5781, pp. 1773-1777.

C.C. Lin, C. Yuan and F.H. Shu, 1969, On the Spiral Structure of Disk Galaxies III. Comparison with Observations, *Ap.J.* **155**, 721-746.

I. Minchev and A.C. Quillen, 2006. *MNRAS* **368**, 623.

P.P. Parenago, 1950, *AZh,* **27**, 150.

A.C. Quillen, 2003. *ApJ* **125**, 785.

D. Russeil, 2003, Star-forming complexes and the spiral structure of our Galaxy, *A.&A.*, **397**, 133-146.

R.S. de Simone, X. Wu & S. Tremaine 2004, *MNRAS*, **350**, 627D.

J. Skuljan, J. B. Hearnshaw and P. L. Cottrell, 1999, *MNRAS,* **308**, 731-740.

In: Galaxies: Evolution, Morphology and Dynamics ISBN: 978-1-62081-185-6
Editors: A. Guerrera Sanchez et al., pp. 165-182

Chapter 7

THE MASS-METALLICITY RELATION: A FOSSIL FROM THE STAR FORMATION HISTORY OF GALAXIES

María Emilia De Rossi *
Consejo Nacional de Investigaciones Científicas y Técnicas, Argentina.
Instituto de Astronomía y Física del Espacio, Argentina.

Abstract

In last years, there has been important progress in the understanding of the chemical evolution of galaxies because of the improvement of observational techniques and the development of more sophisticated numerical codes. In particular, in the Local Universe there is a tight correlation between the stellar mass and metallicity of galactic systems in such a way that brighter and more massive galaxies tend to be more metal-enriched (e.g. Tremonti et al. 2004; Lee et al. 2006). Moreover, this relation seems to evolve with time in the sense that, at a given stellar mass, systems were less enriched in the past (e.g. Erb et al. 2006; Maiolino et al. 2008).

The chemical composition of a galaxy is of fundamental importance to study its formation and evolution as it is the result of the joint action of several physical mechanisms such as the star formation processes, gas infall, Supernovae winds, the initial mass function, stellar nucleosynthesis, among others. Therefore, the determination of correlations between chemical abundances and other properties of galaxies and the study of how these relations evolve with time could provide important clues about possible galaxy formation scenarios (e.g. De Rossi et al. 2007, 2009; Finlator & Davé 2008).

In this work, we will review some of the most important advances made in this sense in the last decades. I will focus mainly on the study of the origin and evolution of the mass-metallicity relation as a relic from the star formation history of galaxies.

Key Words: galaxies: abundances, galaxies: evolution, galaxies: starburst. AMS Subject Classification: 98.62.Bj, 98.62.Lv, 98.62.Ai, 98.62.-g.

*E-mail address: derossi@iafe.uba.ar

1. Introduction

The study of the chemical content of galaxies could provide important clues about the evolutionary history of these systems (e.g. Freeman & Bland-Hawthorn 2002; Pettini 2003, 2006) because their chemical patterns are the result of the joint action of several physical processes such as the star formation histories, chemical enrichment, initial mass function, stellar nucleosynthesis, Supernova (SN) feedback (Larson 1974; Larson & Dinerstein 1975; White & Rees 1978; Kauffmann & Charlot 1998; Somerville & Primack 1999; Cole et al. 2000; Springel & Hernquist 2003), active galactic nuclei feedback (AGN, e.g. Croton et al. 2006), gas infall (Pagel 1997), mergers and interactions (e.g. Pérez et al. 2006; De Lucia & Blaizot 2007; De Rossi et al. 2007, 2009), among others. Therefore, it is plausible to establish constraints on possible galaxy formation scenarios by analysing the evolution of the correlations between the chemical and dynamical properties of galactic systems and contrasting these results with observational ones (e.g. De Rossi et al. 2007; Finlator & Davé 2008).

During the last years, there has been a significant increase in our knowledge about the chemical properties of the Universe because of the improvement of observational techniques which have made possible the exploration not only of nearby galaxies but also the high redshift Universe, as well as the intergalactic (IGM) and intercluster (ICM) medium (De Lucia 2009). Special attention has been given to the study of the fundamental metallicity relations of galactic systems. In particular, the study of the correlation between the stellar mass and metallicity of galaxies (MZR) could reveal important clues for the understanding of the star formation history and chemical enrichment of these systems (e.g. Tissera et al. 2005; De Rossi et al. 2007; Finlator & Davé 2008). The stellar mass of a galaxy constitutes a relic from its star formation history reflecting the total amount of baryons converted into stars and the consequent chemical enrichment of the interstellar medium (ISM). The metallicity, on the other hand, encoded information about the feedback history of the system, modulated by infall and outflows of gas.

In the local Universe, there is a tight correlation between metallicity and luminosity (LZR) in such a way that more luminous systems tend to be more metal enriched (e.g. Lamareille et al. 2004; Saviane et al. 2008). Moreover, this relation seems to evolve with time in the sense that, at a given metallicity, galaxies were brighter the past (e.g. Kobulnicky & Kewley 2004). However, in many studies luminosity has been used as a surrogate for stellar mass because of the difficulties in measuring the later. In fact, the LZR seems to be the result of the more fundamental relation between the metal content of a galaxy and its stellar mass (e.g. Savaglio et al. 2005). The advent of more sophisticated observational techniques in the last decade made possible the study of the local MZR, showing that it follows a very tight correlation (± 0.1 dex) which extends at least ~ 5 orders of magnitude in stellar mass and ~ 2 dex in metallicity (Tremonti et al. 2004; Lee et al. 2006). However, a more detailed analysis indicated that, at a given stellar mass, the MZR seems to depend on the half light radii and specific SFRs. Galaxies with larger half light radii and higher specific SFRs tend to be offset towards lower metallicities (by up to 0.2 dex), effect which could be attributed to differing star formation efficiencies (Ellison et al. 2008). With respect to the redshift evolution, it has been possible to determine that the MZR is also in place at redshifts $z > 0$ in such a way that systems with similar stellar masses were less enriched in the past (e.g. Savaglio et al. 2005; Erb et al. 2006; Maiolino et al. 2008).

In this context, there is an increasing interest in developing galaxy formation models in order to describe in a self-consistent way the variety of complex physical mechanisms that govern the chemical evolution of baryons (e.g. Mosconi et al. 2001; Scannapieco et al. 2005, 2006; Governato et al. 2007) with the aim at reproducing and explaining the observed correlations between galaxy properties. However, the comparison between observations and theory is not straightforward because of the differences in analysis techniques and selection effects, among others.

In this work, I will review some of the most important advances made in this field during the last decade both, from the observational and theoretical point of view. I will focus mainly on the study of the origin and evolution of the MZR and its relation to the star formation history of galaxies.

2. The Observed Luminosity-Metallicity Relation

Lequeux et al. (1979) reported for the first time a correlation between the metallicity and total mass for irregular galaxies, in the sense that more massive galaxies exhibited larger oxygen abundances. However, subsequent studies continued using luminosity as a surrogate for mass, because of the difficulties in measuring the later. Since the pioneer work by Lequeux et al. (1979), the LZR has been widely explored by different research projects which extended it to different Hubble types (e.g. Garnett & Shields 1987; Skillman et al. 1989; Brodie & Huchra 1991; Zaritsky et al. 1994; Richer & McCall 1995; Pilyugin 2001; Garnett et al. 1997; Pilyugin & Ferrini 2000). However, there is observational evidence which suggests that galaxies with different luminosities might not follow a similar LZR. Melbourne & Salzer (2002), for example, analysed a sample of star-forming galaxies detecting that the LZR exhibited a higher slope than that associated to dwarf systems. On the other hand, Cooper et al. (2008) recently studied a sample of 57377 galaxies from the Sloan Digital Sky Survey (SDSS) finding a correlation between metallicity and environment so that more metal-enriched systems are preferentially located in high density regions. These authors claimed that more than 15% of the dispersion observed in the MZR correlates with the density of the environment hosting these galaxies.

Lamareille at al. (2004) determined the LZR in the range $0 < z < 0.15$ finding a linear relation between gas-phase oxygen abundance and the magnitude M_B along the range $-13 < M(b_j) < -22$. As previously reported in the literature, these results indicated that systems with larger luminosities tended to be more chemical-enriched. The same behaviour was observed in the R band of the spectra with a similar rms of ~ 0.27. A simultaneous study of the evolution of the LZR with z suggested that, at a given metallicity, systems were brighter in the past. Nonetheless, as discussed by Lamareille at al. (2004), the interpretation of these trends was affected by selection effects which made difficult to draw a firm conclusion. Firstly, brighter objects tend to be detected a high z, where the sampled volume is large enough to obtain a significant number of these systems However, as the mean luminosity is normalized to the total number of galaxies, this effect cannot explain the evolution of the mean luminosity. On the other hand, fainter galaxies might be found easily at all z, but most of them are observed in the nearby Universe because of the low magnitude limit of the survey. This last issue is more complicated because the statistic of low metallicity galaxies at high z is very poor. In addition, in this work brighter galaxies were excluded from the analysis due to their low signal-to-noise ratio (S/N).

In the literature, different approaches have been proposed in order to explain the origin of the correlation between metallicity and luminosity. In particular, one of the preferred scenarios seems to be the stronger action of SN driven winds in the shallower potential wells of smaller galactic systems, which expel a large amount of metal-enriched material before they have reached a high metallicity. Other possibilities include an increasing efficiency of star formation with stellar mass (e.g. Savaglio et al. 2005) as well as the influence of the infall of less-enriched material (Pagel 1997) on the regulation of the star formation process in galaxies (e.g. Mannucci et al. 2009).

As the gas reservoir is transformed into stars, the chemical content of a galactic system can only grow with time unless it is modulated by an important ejection of metals by strong galactic winds or a significant infall of low-metallicity gas. On its side, the evolution of the luminosity depends on the instantaneous star formation rate (SFR) of the galaxy. Observational evidence indicates that star-forming galaxies are affected by important outflows of gas at galactic scales (e.g. Lehnert & Heckman 1996; Dahlem et al. 1998; Rupke et al. 2002). These outflows are generated by the thermal pressure driven by SN explosions. The influence of SN winds on the ISM of galaxies (commonly regarded as SN feedback), has been reported as an important mechanism which regulates the star formation process in galaxies and has been widely studied during the last decades by resorting to analytic, semi-analytic and numerical modelling (e.g. Larson 1974; Larson & Dinerstein 1975; White & Rees 1978; Kauffmann & Charlot 1998; Somerville & Primack 1999; Cole et al. 2000; Springel & Hernquist 2003; Scannapieco et al. 2006; Dalcanton 2007). It has also been suggested that, in the case that local properties such as SN feedback and the gravitational potential well were the main regulators of the star formation and chemical evolution in galaxies, then the LZR should not depend significantly on the cosmic epoch, as predicted by semi-analytic models (e.g. Somerville & Primack 1999). However, the mean SFR seems to grow until $z \sim 2$ (e.g. Madau et al. 1996), and the early Universe showed a lower level of chemical enrichment. These findings suggest that galactic systems at high redshifts should be brighter than local galaxies of similar metallicities if they have more available gas for the star formation process.

In the last years, with the advent of the telescopes of class 8-10 m and more sophisticated observational techniques, several research programmes have tried to estimate the LZR at different cosmic times and detailed information about a large number of galaxies has been gathered in a wide redshift range (e.g. Pettini et al. 1998, 2001; Kobulnicky et al. 1999; Kobulnicky & Koo 2000; Carollo & Lilly 2001; Hammer et al. 2001; Pettini et al. 2001; Contini et al. 2002; Mehlert et al. 2002; Erb et al. 2003; Kobulnicky et al. 2003; Lemoine-Busserolle et al. 2003; Lilly et al. 2003; Shapley et al. 2004; Kobulnicky & Kewley 2004).

Shapley et al. (2004), for example, studied a sample of 7 star-forming galaxies with $2.1 < z < 2.5$ finding that high redshift galaxies were more luminous that the local counterpart, at a given metallicity. The comparison with different evolutionary models led to the conclusion that these objects were probably the progenitors of metal-rich massive spiral and elliptical systems in the Local Universe.

Pettini et al. (2001), on their side, reported that Lyman Break Galaxies (LBGs) at $z \sim 3$

does not follow the local LZR, lying below the local 12+log(O/H) vs M_B curve. These systems have a much lower metallicity than the expected one given their luminosity, with all of them exhibiting evidence for galactic-scale outflows.

At intermediate redshifts, the results of Kobulnicky & Zaritsky (1999) indicate a no significant evolution of the LZR since $0.11 < z < 0.5$. Moreover, Lilly et al. (2003) obtained a modest evolution of the chemical content of galaxies since $z \sim 0.75$, reporting that the overall metallicity of their sample was only slightly lower by 0.08 ± 0.06 dex than the local counterpart. However, other observational works at similar redshifts suggest a more important offset of the LZR with respect to the local relation. In particular, Kobulnicky et al. (2003) analysed a sample of galaxies at $0.26 < z < 0.82$ and gas-phase abundances of 8.4 <12+log(O/H)< 9.0. They found that, at a given metallicity, high redshift galaxies ($z = 0.6 - 0.82$) were brighter, in mean, by ~ 1 mag than low redshift systems ($z = 0.26 - 0.40$) and by up to ~ 2.4 mag when compared with a local sample ($z < 0.1$). According to this work, the galaxies at higher redshifts exhibited, on average, 40% (0.15 dex) lower metallicities, at a fixed luminosity, than local ones. For these high redshift galaxies ($0.6 < z < 0.8$), the variations with respect of the local LZR were larger for fainter objects ($M_B > -19$) and were very small for brighter systems ($M_B \sim -22$). These results was examined with the PÉGASE code, which allowed the authors to follow the evolution of the luminosity and chemical content of galaxies since their formation towards $z = 0$. In the framework of these models, Kobulnicky et al. (2003) claimed that, in order to explain the slope and evolution of the LZR, at least two of the following mechanisms are valid for star-forming galaxies until $z \sim 0.8$: (1) low luminosity galaxies have lower effective yields than more massive ones, (2) the time scale for the assembly of fainter systems are larger than that associated to more massive ones and (3) less massive galaxies start to assemble later than more massive ones.

On its side, Maier et al. (2004) studied a sample of galaxies at $z \sim 0.4$ and $z \sim 0.64$ with $M_B > -19$, confirming that the LZR is at place at intermediate redshift, but displaced towards lower metallicities and higher luminosities when compared to the local reference. A possible scenario of evolution was suggested in which galaxies at intermediate redshift suffer a decrease in their luminosities by $\sim 0.5 - 0.9$ mag in ~ 6 Gyr, due to the decrease in their SFRs, and an increase in metallicity by ~ 0.3 dex.

By analysing a sample of galaxies at $0.3 < z < 1.0$, Kobulnicky & Kewley (2004) also found evidence for a significant evolution of the LZR at medium redshifts. They found that, at $z \sim 1$, galaxies with $-18.5 < M_B < -21.5$ were displaced towards lower metallicities by (0.14 ± 0.05) dex with respect to the local counterpart. According to this work, the observed rate of chemical enrichment imply that $(28.00 \pm 0.07)\%$ of the metals in local galaxies has been processed since $z = 1$, consistently with the claim that $\sim 38\%$ of the stars in the Universe have been formed during the same time interval. Moreover, the authors detected that the slope of the LZR exhibited a dependance on redshift in such a way that, at a given luminosity, the metallicities associated to fainter objects suffered a more significant increase with time. These trends are in agreement with the downsizing scenario for star formation, where the most massive galaxies form first and galaxy formation proceeds from larger to smaller mass scales (Cowie et al. 1996).

3. The Observed Stellar Mass-Metallity Relation

In the context of the study of galaxy evolution, the stellar mass is a more fundamental parameter than luminosity because it is more closely linked to the dynamical properties of the system. Although the luminosity is mainly regulated by the dynamical evolution of galaxies, it is also affected by other physical processes such as dust attenuation, which could generate different kind of selection effects and systematics uncertainties. In particular, one of the main difficulties is related to the fact that the absolute magnitude of a galaxy depends on the chemical abundances and the ages of its stellar populations so that it is not straightforward to determine whether a system is young and metal-enriched or old and metal-poor (the so called age-metallicity degeneracy).

The stellar mass and the metallicity are considered fundamental properties of galactic systems. The former represents the amount of gas converted into stars, while the later is an indicator of the amount of gas synthesised in stellar interiors and regulated by the exchange of mass with the IGM. Therefore, the study of the evolution of these parameters with time and how they correlate with each other is crucial to understand the physical processes that govern the star formation efficiency in galaxies.

In the Local Universe, Tremonti et al. (2004) determined the MZR by analysing a sample of ~ 53000 star-forming galaxies extracted from the SDSS. These authors worked in the range $0.005 < z < 0.30$, obtaining a tight (± 0.1 dex) correlation between stellar mass and gas-phase oxygen abundance, which extended over 3 decades in stellar mass and approximately a factor of 10 in metallicity. The correlation exhibited a linear trend from $\sim 10^{8.5} M_{\odot}$ towards $\sim 10^{10.5} M_{\odot}$, with a gradual flattening for larger stellar masses. This variation of the slope with mass was attributed to the more efficient action of SN winds in systems with smaller dynamical masses. The main difference between the LZR and the MZR derived in this work was the more pronounced turnover at the high metallicity end when stellar mass was taken as the independent variable. One possible interpretation for the observed behaviour could be related to the fact that smaller galaxies tend to have low mass-to-light ratios (M/L) in comparison with the M/L of more massive systems. Hence, at a given metallicity, less massive objects tend to be displaced along the horizontal axes in a different fashion than more massive ones. In fact, Tremonti et al. (2004) showed that M/L and dust variations tend to smear out the turnover of the LZR with respect to the MZR.

Gallazzi et al. (2005) also detected a gradual flattening above $\sim 3 \times 10^{10} M_{\odot}$ for galaxies in the SDSS when studying the dependance of stellar metallicity and light weighted ages as a function of stellar mass. In particular, these results indicated that smaller galaxies tend to be dominated by young stellar populations with low metallicities. Interestingly, the characteristic stellar mass where the turnover occurs is similar to the transition mass identified by Kauffmann et al. (2003) for a sample of galaxies from the SDSS, which separates two galaxy populations with different physical properties. According to Kauffmann et al. (2003), galaxies with stellar masses smaller than $\sim 3 \times 10^{10} M_{\odot}$ have low surface mass densities, low concentration indices typical of discs and young stellar populations, while larger systems exhibit high surface mass densities, high concentrations typical of bulges and old stellar populations. However, the large scatter found by Gallazzi et al. (2005) in the correlations between stellar mass, metallicity and stellar age made these authors to suggest that the physical properties of galaxies in the Local Universe seem not to be univocally determined by stellar mass. Other processes, such as infall or outflows, might have played a significant role on the star formation history and chemical enrichment of these systems. Furthermore, as we have already mentioned, environment seems also to constitute a significant source of scatter in the MZR. Cooper et al. (2008) found that 15% of the measured scatter in the local MZR correlates with environment in the sense that more metal-enriched galaxies tend to inhabit high density regions such as groups or clusters, where the local environment might influence the action of feedback and gas infall relative to field galaxies of similar masses.

In the simple close-box model, the chemical evolution progresses as the gas reservoir is converted into stars inside galactic systems, without the influence of infall or outflows, and the metallicity Z [1] is given by

$$Z = y \ln(1/\mu) \tag{1}$$

where μ is the gas fraction and y corresponds to the true yield, which depends on stellar nucleosynthesis and is assumed to be constant (Garnett 2002). The true yield represents the ratio of the mass of metals produced and released as a by-product of star formation to the baryonic mass locked in long-lived stars and remnants. Equation 1 can be inverted to obtain the effective yield $y_{\rm eff} = Z/\ln(1/\mu)$. If the simple close box model is valid, then $y_{\rm eff}$ does not depend on the mass of the system and is equal to y. Any deviation from this behaviour which leads to a decrease in $y_{\rm eff}$ from the true yield or as a function of mass implies that the system is subject either to important outflows of enriched material or infall of metal-poor gas (see Dalcanton 2007 for a more detailed insight into the problem). There is no plausible mechanism which could lead to an increase in the effective yield with respect to the close-box value (Edmunds 1990; Köppen & Edmunds 1999), with the only exception of the unlikely accretion of metal-rich gas.

Tremonti et al. (2004) analysed the dependance of the effective yields with baryonic mass for their sample of galaxies, obtaining a decrease by a factor of 10 from the most massive systems to dwarfs. Following Larson (1974), the authors interpreted these trends as the preferential loss of metals from the shallower potential wells of less massive objects driven by ubiquitous galactic winds in star-forming galaxies.

Taking advantage of the sensitivity of the Spitzer Space Telescope, Lee et al. (2006) were able to determine the MZR for a sample of 27 dwarf irregular galaxies in the Local Universe, finding a strong correlation which extended that reported by Tremonti et al. (2004) by 2.5 dex towards lower stellar masses. These findings showed that the MZR seems to be a well-defined tight correlation at least in a range of 5 order of magnitude in stellar mass and 2 dex in metallicity. Moreover, the MZR exhibited a similar dispersion (≈ 0.1 dex) for the whole range of observed stellar masses. These results together with the larger effective yields obtained for some dwarfs (which, in some cases, exceeded the asymptotic value for larger systems in the SDSS) leaded Lee et al. (2006) to conclude that these trends were difficult to be explained by a model in which galactic superwinds or outflows are responsible for the low metallicities exhibited by fainter or less massive galaxies. They suggested instead a scenario in which metals might be lost by a low-energetic mechanism than blowouts. A low efficiency of star formation in dwarf irregular galaxies was proposed

[1] Z is calculated as the ratio of the mass of all elements heavier than helium to the total mass in a given baryonic phase.

as a plausible solution.

With the advent of more sophisticated surveys of galaxies at intermediate and high redshifts, it has recently become possible to estimate the chemical enrichment of baryons at $z > 0$, getting deeper insight into the high redshift Universe. In particular, Savaglio et al. (2005) studied the MZR for a sample of galaxies from the Gemini Deep Deep Survey and the Canada-France Redshift Survey at $0.4 < z < 1.0$, detecting for the first time a tight correlation between mass and metallicity at intermediate redshifts. These authors found that the MZR exhibits a stronger and tighter correlation than the LZR, which suggests the more fundamental nature of the former. In addition, they obtained an evolution of the MZR in the sense that, at a given stellar mass (M_*), galaxies at $z \sim 0.7$ have metallicities lower by ~ 0.15 dex than those corresponding to the local sample studied by Tremonti et al. (2004). These trends are consistent with those recently found by Lara-López et al. (2009), who reported a decrease in metallicity by ~ 0.1 dex, on average, between $z = 0$ and $z = 0.4$ for a sample of massive ($M_* > 10^{10.5} M_\odot$) star-forming galaxies from the SDSS-data release 5 (SDSS-DR5). Liang et al. (2006) also found a strong evolution of the MZR in a similar redshift range however, due to discrepancies between the analysis methods (see Liang et al. 2006 for details), their results cannot be directly contrasted with the ones derived by Savaglio et al. (2005).

By comparing their sample with that derived by Tremonti et al. (2004) at $z \sim 0.1$ and a small number of Lyman-Break galaxies at $z \sim 2.3$, Savaglio et al. (2005) obtained that the time scale of evolution in stellar mass and chemical enrichment tend to increase for low mass systems. In particular, the authors showed that the observed evolution of the MZR may be well-described by a closed-box model by assuming a shorter e-folding time for star formation in more massive galaxies, in agreement with the downsizing scenario for galaxy formation. However, the authors acknowledged the fact that mergers and interactions between galaxies, which are not consider in this simple model, might have also played an important role on the star formation history and chemical evolution of galactic systems. Nevertheless, it is worth noting that similar trends were reported by Juneau et al. (2005) by studying a sample from the Gemini Deep Deep Survey at $0.8 < z < 2$. According to this work, the star formation history of massive galaxies ($M_* > 10^{10.8} M_\odot$) is nearly completed by $z \sim 1.5 - 2$, intermediate-mass objects ($10^{10.2} M_\odot \leq M_* \leq 10^{10.8} M_\odot$) have already formed their dominant stellar mass component at $z \sim 1$, while smaller galaxies seem not to have stopped their star formation at least for the redshift range covered by this survey.

An alternative scenario to explain the evolution of the MZR at intermediate redshifts was suggested by Lamareille et al. (2009) who, contrary to Savaglio et al. (2005), found a flatter slope of this relation compared with the local one. They obtained that galaxies with $\sim 10^{9.4} M_\odot$ exhibited -0.18 dex lower chemical abundances at $z \sim 0.77$ when compared with the local sample. On the other hand, systems with $\sim 10^{10.2} M_\odot$ were less enriched by -0.28 dex at ~ 0.77 with respect to local galaxies with similar stellar masses. They obtained that, by assuming the close-box hypothesis, the observed evolution of the MZR could only be reproduced if galaxies had similar star formations at any mass, which was in clear contradiction with observations. They proposed, instead, an 'open-closed' model in which low mass systems evolve as open-boxes losing metals through galactic outflows, while massive galaxies evolve as closed-boxes retaining their gas-phase component due to their more important gravitational potential wells. In this context, it was suggested that the hierarchical aggregation of the structure plays an important role regulating the potential well of galaxies by merging and accretion processes. This work was extended towards $z \sim 1.24$ by Pérez-Montero et al. (2009), finding that galaxies at $z \sim 1$ exhibit metallicities ~ 0.3 dex smaller than local ones, in agreement with other works (Shapley et al. 2005; Cowie & Barger 2008; Liu et al. 2008). However, they did not detect a significant evolution of the MZR from $z \approx 1.0$ to $z \approx 0.7$.

The redshift range $1 < z < 2$ has been regarded as a fundamental epoch for the star formation history and the chemical evolution of the Universe as a whole, hence, the determination of the shape and evolution of the MZR at this epoch could provide important information for models of galaxy formation. By analysing a sample of 20 star-forming galaxies at $1.0 < z < 1.5$, Liu et al. (2008) found that the correlation between gas-phase oxygen abundance and stellar mass was already present at this cosmic time, but showing an offset towards lower abundances with respect to the local MZR: low mass galaxies ($M_* \sim 8 \times 10^9 M_\odot$) were displaced by $\sim 0.2(0.35)$ dex with respect to the local relation according to the N2 (O3N2) indicator, while more massive systems ($M_* \sim 5 \times 10^{10} M_\odot$) showed a change by at least ~ 0.2 dex in their metal content, according to the O3N2 indicator. These results are consistent with a recent work by Queyrel et al. (2009), who studied a sample of nine star-forming galaxies from the VIMOS VLT Deep Survey at $z \sim 1.4$, finding that these galaxies are displaced by $\sim 0.2 - 0.4$ dex towards lower abundances with respect to the local MZR derived from the SDSS. However, the correlation detected between metallicity and stellar mass was very week, with only a small tendency of larger systems to be more metal-enriched. Thus, it was not possible to draw a firm conclusion. However, it is encouraging that, when completed, the MASSIVE (Mass Assembly Survey with SINFONI in VVDS) survey will make possible to study the properties of around ~ 80 galaxies within $1 < z < 2$ and determine fundamental relations between them such as the MZR.

Getting deeper insight into the high redshift universe, Erb at al. (2006) analysed a sample of 87 star-forming galaxies at $<z>= 2.26 \pm 0.17$, finding that the MZR is also present at this redshift but displaced by ~ 0.3 dex towards lower abundances than the relation reported by Tremonti et al. (2004) in the Local Universe. Interestingly, contrary to the local trend, high redshift galaxies showed slightly lower effective yields for larger baryonic masses. In particular, these authors detected a correlation between metallicity and gas fraction which could be explained by a model with supersolar yields ($y \sim 1.5 Z_\odot$) and an outflow rate ~ 4 times higher than the SFR. In this model, the chemical evolution of galaxies is driven by the star formation process and is likely regulated by strong metal-loaded outflows in galaxies of all masses. This picture is supported from observations at low and high redshifts which suggest that the outflow rate is comparable to the SFR or even higher (e.g. Martin 1999, 2003; Pettini et al. 2000). As discussed by Erb et al. (2006), in this simple model it is assumed that the metallicity of the wind is similar to the one observed in the galaxy, which is not necessary true. Thus, it may be possible to fit the data by developing a new model with metal-enhanced outflows and lower outflows rates. With respect to the gas fractions, these authors found an average value (~ 0.5) 2 times greater than the observed by Tremonti et al. (2004) (~ 0.2), indicating that systems at $z \sim 2$ are at an earlier evolutionary stage, which is consistent with their lower level of chemical enrichment. By analysing the cluster properties of the galaxies in their sample, Erb et al. (2006) concluded that these systems could likely be associated to the progenitors of early-type galaxies in the SDSS, while the star-forming galaxies studied by Tremonti et al. (2004) are too weakly clustered to be the descendant of the $z \sim 2$ population.

In a recent work, Maiolino et al. (2008) reported the first results from the AMAZE project, which is focused on measuring the evolution of the MZR at $z > 3$ by using spectroscopic techniques in the near infrared. Comparing these results with the ones obtained from previous surveys [2] , the MZR at $z \sim 3.5$ shows a stronger evolution at $2.2 < z < 3.5$ than the appreciated at lower redshifts ($z < 2.2$). A significant evolution was detected also for more massive systems, suggesting that $z \sim 3$ is a crucial epoch for the star formation history and chemical evolution of the Universe. Nevertheless, Maiolino et al. (2008) found evidence of a more important chemical evolution in the case of less massive galactic systems which, according to these authors, could be considered as the chemical version of the galaxy downsizing: the time scale for chemical enrichment seems to be larger in smaller systems, while more massive ones reach higher metallicities earlier. Maiolino et al. (2008) proposed a scenario in which massive galaxies at $z > 3$ are formed by the hierarchical aggregation of unevolved systems, whose star formation efficiencies are low. Most of the star formation and chemical enrichment proceeds once these substructures are assembled onto larger ones. In this context, the energy injected into the ISM by SN feedback might provide a suitable mechanism to avoid higher star formation efficiencies in low mass substructures.

Another encouraging project, aimed at study LBGs at $z \sim 3$, is 'Lyman-break galaxies Stellar populations and Dynamics' (LSD). The first results were presented by Mannucci et al. (2009), reporting a strong evolution of the MZR up to these redshifts, with LBGs tending to have metallicities ranging from 0.1 and 0.5 solar. By analysing the gas fractions and chemical-enrichment, they obtained that these systems are massive but low metal-enriched, indicating that mass assembly could have occurred before the bulk of the star formation. In agreement with previous works at $z > 0$ (e.g. Erb et al. 2006), they found larger effective yields for smaller galaxies, reaching solar values at the low mass end. They suggested that this behaviour could be explained by appealing to gas infall with rates of the order of the SFRs or merging with gas-rich metal-poor galaxies, as a consequence of which starbursts occur. Outflows are probably present but they seem not to dominate the chemical evolution at this cosmic epoch.

4. Predictions from Theoretical Models

In the light of these observational findings, it is of great relevance the development of theoretical models that can describe the several physical mechanisms responsible for the chemical evolution and star formation history of galaxies (e.g. SN feedback, gas cooling, mergers and interactions, stellar nucleosynthesis, gas infall, among others). These models should be able to reproduce in a consistent fashion the fundamental relations observed between the dynamical and chemical properties of galaxies and their evolution with cosmic epoch.

An insight into the problem of the MZR by resorting to numerical models was provided by Tissera et al. (2005) and De Rossi et al. (2007). These authors studied the origin and evolution of the MZR and LZR by employing numerical hydrodynamical simulations

[2]Maolino et al. (2008) have recalibrated all the metallicity and mass indicators in previous works in order to match their own calibrations.

within a cosmological framework. They found that the shape of the local simulated MZR fitted prettily well the observed behaviour, suggesting that the hierarchical aggregation of the structure played a fundamental role on its determination. They detected a characteristic stellar mass at $\sim 3 \times 10^{10} M_{\odot}$ which segregates systems dominated, on average, by old, red and chemical-enriched stellar populations from those with larger gas fractions and younger stars. As discussed by De Rossi et al. (2007), the characteristic mass arises as a consequence of the hierarchical aggregation process in a ΛCDM cosmology. Massive galaxies form by the fusion of smaller galactic systems dominated by old stellar populations with no significant star formation after the merger event. Hence, the stellar mass of the remnant will be larger than that associated with each progenitor, while the metallicity will remain almost the same originating the flattening of the MZR, consistently with the observational trend. On the other hand, this model predicts that smaller systems are not significantly affected by galaxy interactions or merger events during their evolution. Therefore, the star formation inside them proceeds in a more passive way resulting in a stronger correlation between stellar mass and chemical enrichment. With respect to the evolution of the MZR, De Rossi et al. (2007) predicted a weaker evolution by ~ 0.10 dex in the range $0 < z < 2$ than that expected from observations. On the other hand, at $z \sim 0$, they obtained larger effective yields for systems with smaller baryonic masses, contrary to observations (e.g. Tremonti et al. 2004). Thus the authors suggested the need for the implementation of strong SN feedback in their model. This issue was tackled by De Rossi & Tissera (2010), who showed that a suitable treatment of metal-loaded galactic outflows (Scannapieco et al. 2006) would be able to solve the problem of the slope of the yield-mass relation in the Local Universe. However, at a given stellar mass, the effective yields derived from this model are lower than the observed ones. With respect to the evolution of the MZR, the simulations were able to reproduce the MZR reported by Maiolino et al. (2008) at $z \sim 3.5$, but they predicted a weaker evolution towards $z \sim 0$ than that reported in observational works. The authors concluded that the SN feedback efficiency needed to be revised by varying the parameters of the SN model (see also De Rossi et al. 2010, in preparation).

The modelization of the injection of energy into the ISM by SN explosions as well as the chemical feedback is a complicated task due to the lack of a clear understanding of the physics involved and numerical difficulties to describe the wide range of associated physical scales. In the scientific literature, there are different approaches to tackle these issues. First results, obtained by Brooks et al. (2007) by performing a physically motivated description of SN feedback and chemical enrichment in cosmological simulations, showed the important role of galactic winds and lower star formation efficiencies in dwarf galaxies as regulators of the functional form of the MZR. This model was able to described the shape and normalization of the MZR at $z = 0$ and $z = 2$. The implementation of SN feedback was crucial to reduce the SF efficiencies in low mass galaxies and originating the observed turnover in the MZR. Without including a SN feedback mechanism, the star formation proceeds more rapidly generating a more important chemical enrichment in low mass systems leading to a flatter MZR. On their side, by performing cosmological simulations which include treatments for SN feedback, Kobayashi et al. (2007) obtained that wind fraction correlated with stellar metallicities suggesting that the origin of the MZR is from galactic winds. Another interesting approach was introduced by Finlator & Davé (2008), who studied the influence of metal-enriched outflows on the chemical enrichment of galaxies by using numerical simulations. By comparing a model without outflows with a constant-wind model and a momentum-driven model, they concluded that this last scenario is able to explained the observed slope, normalization and scatter of the MZR at $z \sim 2$. Moreover, this model was able to better describe the chemical abundances of the IGM at $z > 2$. In this framework, it was suggested that the MZR is governed by an evolving equilibrium between the enrichment rate as a consequence of star formation and a dilution rate due to gas infall, while outflows influence the MZR by limiting the gas available for star formation. Mouchine et al. (2008) investigated also the MZR by performing numerical hydrodynamical simulations in a Cold Dark Matter Universe, which incorporate treatments for star formation, metal-dependent cooling, SN feedback and a chemical enrichment of the ISM that relaxed the instantaneous recycling approximation. The derived local MZR showed general agreement with observations. According to these authors, an increase of the star formation efficiency with mass and the different assembly histories of galaxies could explain the origin of the shape and scatter of the MZR in their simulations. However, contrary to observations, they obtained no correlation between age and stellar mass. Smaller systems at $z = 0$ exhibited older ages that the expected from observations and more massive ones did not show a significant evolution, leading the authors to suggest that their feedback model might be incomplete or other feedback processes would need to be taken into account.

An alternative and innovative approach to explain the origin of the MZR was suggested by Köppen et al. (2007), which is based on a variable integrated initial mass function. With this implementation, the effects of galactic winds on chemical evolution need not to be so strong. As explained by Köppen et al. (2007), the idea is that galaxies with low SFRs tend to have clusters up to a lower mass limit, with less massive clusters exhibiting smaller probabilities of hosting very massive stars. Then, the range of stellar masses covered by the IMF has a lower upper limit during epochs of lower star formation activity. In the model, the main effect is the lowering of the effective upper mass limit of stars in regions with low SFRs, which leads to a decrease in the number of SNs of type II. This model is able to describe the MZR in good agreement with observational data without resorting to galactic outflows, offering another plausible mechanism which could account for the origin of the MZR in galaxies.

In the context of chemical analytical models, in a recent work, Calura et al. (2009) studied the evolution of the MZR for galaxies with different morphology types: ellipticals, spirals and irregulars. They reported that the MZR at $z \sim 3.5$ seems to be constituted mainly by proto-ellipticals, while at $z \sim 2.2$ there would be a mix of proto-spirals and proto-ellipticals and, at $z \sim 0$, the MZR is reproduced by considering both spirals and irregulars. They showed that, in their models, the MZR has its origin mainly in the lower star formation efficiencies exhibited by smaller galaxies, without the need to resort to the implementation of a more efficient mass-loss in low-mass systems.

5. Conclusion

In this chapter, I have summarized some of the most important advances made during the last decades regarding the study of the LZR and MZR since the pioneer work by Lequeux et al. (1979). It is clear that, in the Local Universe, brighter and more massive galaxies tend to be more metal-enriched (e.g. Tremonti et al. 2004). Furthermore, these relations tend to evolve with time in such a way that, at a given stellar mass or luminosity, systems exhibit a lower level of chemical enrichment in the past (e.g. Maiolino et al. 2008).

Although it seems to be a general consensus about the presence of these relations, both in the nearby and the distant Universe, there are still several open problems concerning the origin of the shape of the relation, its scatter and evolution, which could only be addressed with the progress of the observational techniques and the development of more sophisticated theoretical models.

From the observational point of view, one of the most challenging task is the determination of the absolute metallicities of galactic systems, as the use of different metallicity calibrators yields different results. This issue has been discussed in detail by Kewley & Ellison (2008), who studied the influence of 10 different metallicity indicators on the MZR. They found that the use of different calibrators significantly alter the shape and offset (by up to 0.7 dex) of the MZR, showing the relevance of converting metallicities to the same indicator when comparing samples corresponding to different works. In addition, according to these authors, the turnover of the MZR at $M_* \sim 10^{10}M_\odot$ is affected by the aperture covering fraction. This issue complicates the comparison between surveys at different redshifts (e.g. Maiolino et al. 2008) and, hence, makes difficult a reliable determination of the evolution of the MZR. Moreover, at the same redshift, the different aperture effects for galaxies with different masses could also modify the shape of the relations. However, as discussed by Maiolino et al. (2008), these effects are weaker than the observed evolution at least in the range $0.0 < z < 3.5$. These results indicate that there is a real correlation between the stellar mass of galaxies and their chemical content, and that it evolves with time in such a way that systems with similar masses exhibit lower metallicities in the past. However, there are still controversies about the level of the evolution and the absolute values of metallicities which will be only possible to be solved with the progress of observational techniques.

In this context, the challenge for theoretical models will be providing a suitable framework to reproduce and explain the origin of the observed correlation between galaxy properties. As we have seen, different approaches have been proposed in the literature to address the problem of the origin of the MZR. Most of them are based in the analyses of deviations from the close-box model. The preferred scenarios seem to be an increasing efficiency of star formation for larger systems or the more efficient action of galactic winds in low mass galaxies. However, many other physical phenomena are involved in the evolution of galaxies which cannot be discarded: gas infall, environment, mergers and interactions, gas accretion, initial mass function, among others. The complexity of these processes and the different physical scales associated with them makes it necessary to resort to numerical methods in order to tackle the problem in a self-consistent way.

In this context, an important progress has been done during the last years by using analytical, semi-analytical and numerical models of galaxy formation (e.g. De Lucia et al. 2004; Scannapieco et al. 2006). It has been shown that mechanisms such a SN feedback and mergers and interactions seem to have played a fundamental role in the evolution of galaxies and, in particular, the MZR. However, there are still open problems (Tissera et al. 2010). At this moment, one of the most challenging one is the difficulty to describe the wide range of physical scales associated to galaxy evolution. The advent of new technologies together with the progress in numerical techniques and the development of more sophisticated theoretical models will probably contribute a great deal to overcome these issues in the near future.

Acknowledgements

I would like to thank Nova Science Publishers, Inc. for having invited me to participate in their publishing program and the reviewer Dr. Patricia B. Tissera for a careful reading of this chapter and many useful suggestions which helped to improve this manuscript. I am also grateful to CONICET, PICT 32342 (2005) and PICT 245 (2006).

References

[1] J. P. Brodie and J. P. Huchra, *ApJ* 1991, 379, 157-167

[2] A. M. Brooks, F.Governato, C. M. Booth, B. Willman, J. P. Gardner, J. Wadsley, G. Stinson and T. Quinn, *ApJ* 2007, 655, L17-L20

[3] F. Calura, A. Pipino,C. Chiappini, F. Matteucci and R. Maiolino, *A&A* 2009, 504, 373-388

[4] C. M. Carollo and S. J. Lilly, *ApJ* 2001, 548, L153-L157

[5] S. Cole,C. G. Lacey, C. M. Baugh and C. S. Frenk, *MNRAS* 2000, 319, 168-204

[6] T. Contini, M. A. Treyer, M. Sullivan and R. S. Ellis, *MNRAS* 2002, 330, 75-91

[7] M. C. Cooper, C. A. Tremonti, J. A. Newman and A. I. Zabludoff, *MNRAS* 2008, 390, 245-256

[8] L. L. Cowie, A. Songaila, E. M. Hu, J. G. Cohen, *AJ* 1996, 112, 839-864

[9] L. L. Cowie and A. J. Barger, *ApJ* 2008, 686, 72-116

[10] D. J. Croton et al. *MNRAS* 2006, 365, 11-28

[11] M. Dahlem, K. A. Weaver and T. M. Heckman, *ApJS* 1998, 118, 401-453

[12] J. J. Dalcanton, *ApJ* 2007, 658, 941-959

[13] G. De Lucia, G. Kauffmann, S. D. M. White, *MNRAS* 2004, 349, 1101-1116

[14] G. De Lucia and J. Blaizot, *MNRAS* 2007, 375, 2-14

[15] G. De Lucia, *arXiv* 2009, arXiv:0912.1862

[16] M. E. de Rossi, P. B. Tissera and C. Scannapieco, *MNRAS* 2007, 374, 323-336

[17] M. E. de Rossi, P. B. Tissera, G. De Lucia and G. Kauffmann, *MNRAS* 2009, 395, 210-217

[18] M. E. de Rossi and P. B. Tissera, 2010, to appear in the Proccedings of the the IAU's XXVII GENERAL ASSEMBLY, IAU Symposium No. 262

[19] M. G. Edmunds, *MNRAS* 1990, 246, 678-685

[20] S. L. Ellison, D. R. Patton, L. Simard and A. W. McConnachie, *ApJ* 2008, 672, L107-L110

[21] D. K. Erb, A. E. Shapley, C. C. Steidel, M. Pettini, K. L. Adelberger,M. P. Hunt and A. F. M. Moorwood, Cuby, *J.-G. ApJ* 2003, 591, 101-118

[22] D. K. Erb, A. E. Shapley, M. Pettini,C. C. Steidel,N. A. Reddy, K. L. Adelberger, *ApJ* 2006, 644, 813-828

[23] K. Finlator and R. Davé, *MNRAS* 2008, 385, 2181-2204

[24] K. Freeman and J. Bland-Hawthorn, *ARA&A* 2002, 40, 487-537

[25] A. Gallazzi, S. Charlot,J. Brinchmann, S. D. M. White and C. A. Tremonti, *MNRAS* 2005, 362, 41-58

[26] D. R. Garnett and G. A. Shields, *ApJ* 1987, 317, 82-101

[27] D. R. Garnett,G. A. Shields,E. D. Skillman,S. P. Sagan and R. J. Dufour, *ApJ* 1997, 489, 63-86

[28] D. R. Garnett, *ApJ* 2002, 581, 1019-1031

[29] F. Governato,B. Willman, L. Mayer, A. Brooks, G. Stinson, O. Valenzuela, J. Wadsley and T. Quinn, *MNRAS* 2007, 374, 1479-1494

[30] F. Hammer, N. Gruel,T. X. Thuan,H. Flores, L. Infante, *ApJ* 2001, 550, 570-584

[31] S. Juneau et al. *ApJ* 2005, 619, L135-L138

[32] G. Kauffmann and S. Charlot, *MNRAS* 1998, 294, 705-717

[33] L. J. Kewley and S. L. Ellison, *ApJ*, 681, 1183-1204

[34] C. Kobayashi, V. Springel, S. D. M. White, *MNRAS* 2007, 376, 1465-1479

[35] H. A. Kobulnicky ,R. C. Kennicutt and J. L. Pizagno, *ApJ* 1999, 514, 544-557

[36] H. A. Kobulnicky and D. Zaritsky, *ApJ* 1999, 511, 118-135

[37] H. A. Kobulnicky and D. C. Koo, *ApJ* 2000, 545, 712-727

[38] H. A. Kobulnicky et al. *ApJ* 2003, 599, 1006-1030

[39] H. A. Kobulnicky and L. J. Kewley, *ApJ* 2004, 617, 240-261

[40] J. Köppen and M. G. Edmunds, *MNRAS* 1999, 306, 317-326

[41] J. Köppen,C. Weidner and P. Kroupa, *MNRAS* 2007, 375, 673-684

[42] F. Lamareille, M. Mouhcine,T. Contini, I. Lewis and S. Maddox, *MNRAS* 2004, 350, 396-406

[43] F. Lamareille et al. *A&A* 2009, 495, 53-72

[44] M. A. Lara-López,J. Cepa, A.Bongiovanni, A. M. Pérez García, H. Castañeda, M. Fernández Lorenzo and M. Pović,; Sánchez-Portal, M. *A&A* 2009, 505, 529-539

[45] R. B. Larson, *MNRAS* 1974, 169, 229-246

[46] R. B. Larson and H. L. Dinerstein, *PASP* 1975, 87, 911-915

[47] H. Lee, E. D. Skillman, J. M. Cannon, D. C. Jackson, R. D. Gehrz, E. F. Polomski and C. E. Woodward, *ApJ* 2006, 647, 970-983

[48] M. D. Lehnert and T. M. Heckman, *ApJ* 1996, 462, 651-671

[49] M. Lemoine-Busserolle,T. Contini, R. Pelló, J.-F. Le Borgne, J.-P. Kneib and C. Lidman, *A&A* 2003, 397, 839-849

[50] J. Lequeux,M. Peimbert, J. F. Rayo, A. Serrano and S. Torres-Peimbert, *A&A* 1979, 80, 155-166

[51] Y. C. Liang, F. Hammer and H. Flores, *A&A* 2006, 447, 113-119

[52] S. J. Lilly, C. M. Carollo and A. N. Stockton, *ApJ* 2003, 597, 730-750

[53] X. Liu, A. E. Shapley,A. L. Coil, J. Brinchmann, C.-P. Ma, *ApJ* 2008, 678, 758-779

[54] P. Madau, H. C. Ferguson, M. E. Dickinson, M. Giavalisco,C. C. Steidel and A. Fruchter, *MNRAS* 1996, 283, 1388-1404

[55] C. Maier, K. Meisenheimer and H. Hippelein, *A&A* 2004, 418, 475-485

[56] R. Maiolino et al. *A&A* 2008, 488, 463-479

[57] F. Mannucci et al. *MNRAS* 2009, 398, 1915-1931

[58] C. L. Martin, *ApJ* 1999, 513, 156-160

[59] C. L. Martin, *RMxAC* 2003, 17, 56-59

[60] D. Mehlert et al. *A&A* 2002, 393, 809-819

[61] J. Melbourne and J. J. Salzer, *AJ* 2002, 123, 2302-2311

[62] Mosconi, M. B.; Tissera, P. B.,D. G. Lambas and S. A. Cora, *MNRAS* 2001, 325, 34-48

[63] M. Mouchine, B. K. Gibson, A. Renda and D. Kawata, arXiv 2008, arXiv:0801.2476

[64] B. E. J. Pagel, *Nucleosynthesis and chemical evolution of galaxies*; ISBN 0521550610 ;Cambridge University Press: Cambridge, UK, 1997; pp 392.

[65] M. J. Perez,P. B. Tissera, C. Scannapieco, D. G. Lambas, M. E. de Rossi, *A&A* 2006, 459, 361-369

[66] E. Pérez-Montero et al. *A&A* 2009, 495, 73-81

[67] M. Pettini, M. Kellogg, C. C. Steidel, M. Dickinson,K. L. Adelberger, M. Giavalisco, *ApJ* 1998, 508, 539-550

[68] M. Pettini,C. C. Steidel,K. L. Adelberger,M. Dickinson and M. Giavalisco, *ApJ* 2000, 528, 96-107

[69] M. Pettini,A. E. Shapley, C. C. Steidel,J.-G. Cuby, M. Dickinson, A. F. M. Moorwood, K. L. Adelberger and M. Giavalisco, *ApJ* 2001, 554, 981-1000

[70] M. Pettini, In *The Fabulous Destiny of Galaxies: Bridging Past and Present*; Le Brun, V.; Mazure, A.; Arnouts, S.; Burgarella, D.; Ed; ISBN 2914601190; Edition Frontières: Paris, 2006; pp 319

[71] L. S. Pilyugin and F. Ferrini, *A&A* 2000, 358, 72-76

[72] L. S. Pilyugin, *A&A* 2001, 374, 412-420

[73] J. Queyrel et al. *A&A* 2009, 506, 681-687

[74] M. G. Richer and M. L. McCall, *ApJ* 1995, 445, 642-659

[75] D. S. Rupke, S. Veilleux and D. B. Sanders, *ApJ* 2002, 570, 588-609

[76] S. Savaglio et al. *ApJ* 2005, 635, 260-279

[77] I. Saviane, V. D. Ivanov,E. V. Held, D. Alloin,R. M. Rich, F. Bresolin and L. Rizzi, *A&A* 2008, 487, 901-920

[78] C. Scannapieco, P. B. Tissera,S. D. M. White and V. Springel, *MNRAS* 2005, 364, 552-564

[79] C. Scannapieco, P. B. Tissera, S. D. M. White and V. Springel, *MNRAS* 2006, 371, 1125-1139

[80] A. E. Shapley, D. K. Erb, M. Pettini, C. C. Steidel and K. L. Adelberger, *ApJ* 2004, 612, 108-121

[81] A. E. Shapley,A. L. Coil,C.-P. Ma and K. Bundy, *ApJ* 2005, 635, 1006-1021

[82] E. D. Skillman, R. C. Kennicutt and P. W. Hodge, *ApJ* 1989, 347, 875-882

[83] R. S. Somerville and J. R. Primack, *MNRAS* 1999, 310, 1087-1110

[84] V. Springel and L. Hernquist, *MNRAS* 2003, 339, 289-311

[85] P. B. Tissera, M. E. De Rossi and C. Scannapieco, *MNRAS* 2005, 364, L38-L41

[86] P. B. Tissera, 2010, to appear in the Proccedings of the the IAU's XXVII GENERAL ASSEMBLY, IAU Symposium No. 262

[87] C. A. Tremonti, et al. *ApJ* 2004, 613, 898-913

[88] S. D. M. White and M. J. Rees, *MNRAS* 1978, 183, 341-358

[89] D. Zaritsky,R. C., Jr. Kennicutt, J. P. Huchra, *ApJ* 1994, 420, 87-109

In: Galaxies: Evolution, Morphology and Dynamics
Editors: A. Guerrera Sanchez et al., pp. 183-195
ISBN: 978-1-62081-185-6

Chapter 8

TURBULENT FORMATION OF PROTOGALAXIES AT THE END OF THE PLASMA EPOCH: THEORY AND OBSERVATIONS

Rudolph E. Schild[1,*] and Carl H. Gibson[2,†]
[1]Center for Astrophysics, 60 Garden Street, Cambridge, MA, US
[2]University of California San Diego, La Jolla, CA, US

Abstract

The standard model of gravitational structure formation is based on the Jeans 1902 acoustic theory, neglecting nonlinear instabilities controlled by viscosity, turbulence and diffusion. A linear instability length scale equal to the sound speed times the gravitational freefall time emerges. Because the Jeans scale LJ for the hot primordial plasma is always slightly larger than the scale of causal connection ct, where c is the speed of light and t is the time after the big bang, it has been assumed that no plasma structures could form without guidance from a cold (so LJ CDM is small) collisionless cold-dark-matter CDM fluid to give condensations that gravitationally collect the plasma. Galaxies by this CDM model are therefore produced by gradual hierarchical-clustering of CDM halos to galaxy mass over billions of years, contrary to observations showing that well formed galaxies existed much earlier. No observations exist of CDM halos. However, Gravitational instability is non-linear and absolute, controlled by viscous and turbulent forces or by diffusivity at Schwarz length scales smaller than ct. Because the universe during the plasma epoch is rapidly expanding, the first structures formed were at density minima by fragmentation when the viscous-gravitional scale LSV first matched ct at 30,000 years to produce protosupercluster voids and protosuperclusters. Weak turbulence produced at expanding void boundaries guide the morphology of smaller fragments down to protogalaxy size just before transition to gas at 300,000 years. The size of the protogalaxies reflect the plasma Kolmogorov scale with a linear and spiral morphology predicted by the Nomura direct numerical simulations and confirmed by deep field space telescope observations. On transition to gas the kinematic viscosity decreases by ten trillion so the protogalaxies fragment at Jeans scale clouds, each with a trillion earth-mass planets, as predicted by Gibson 1996 and observed by Schild 1996. The planets promptly form stars near the cores of the protogalaxies, but as the universe cools the

* E-mail address: rschild@cfa.harvard.edu.
† E-mail address: cgibson@ucsd.edu, http://sdcc3.ucsd.edu/~ir118.

planets gradually freeze to form the galaxy dark matter. Space telescope observations of the most distant galaxies confirm the Kolmogorov scale size and Nomura geometry of the protogalaxies. High resolution images of planetary nebula and supernova remnants reveal thousands of frozen hydrogen-helium dark matter planets with large atmospheres evaporated by such events. Galaxy mergers show frictional trails of young globular clusters formed in place, proving that dark matter halos of galaxies consist of dark matter planets in metastable clumps. Galaxy evolution is guided by friction from this galaxy dark matter.

Introduction

The standard model of gravitational structure formation and the formation of galaxies is based on a linear acoustic instability theory proposed by James Hopwood Jeans in 1902 in a lengthy monograph titled "The stability of a spherical nebula" transmitted to the Philosophical Transactions of the Royal Society of London Series A (Vol. 1999, pp. 1-53) through G. H. Darwin, son of Charles Darwin. The purpose of the work was to understand how the solar system might have formed from a gas cloud. Galaxies and the expansion of the universe were not known at that time. The alternative theory due to G. H. Darwin in his 1889 paper (Vol. 180, pp. 1-69) was titled "On the mechanical conditions of a swarm of meteorites, and on theories of cosmogony". Neither of the theories are correct as a description of how either the solar system or galaxies were formed. Darwin's theory focusing on meteorites [1] is a better guide than Jeans' in both cases since accretion within clumps of primordial planets and effective viscosity from planet collisions are crucial and the collapse of gas clouds following Jeans 1902 [2] is irrelevant. The Jeans acoustic scale has only one known application, which is to set the mass of globular star clusters at the plasma to gas transition (decoupling), but it is for different reasons than described by Jeans. Modern fluid mechanical concepts [3-19] applied to cosmology are termed hydro-gravitational-dynamics HGD. Relevant length scales are summarized in Table 1.

The fluid mechanics of Jeans 1902 is that of the 19th century. Jeans started with the Euler equations that neglect viscous forces, used linear perturbations stability analysis that neglects turbulence forces, and took no account of diffusivity effects arising from weakly collisional non-baryonic dark matter that probably constitutes about 97% of the rest mass of the universe where the universe is flat with density approximately equal to the present critical density of $\rho_{Crit} = 10^{-26}$ kg m^{-3} with no dark energy or cosmological constant Λ. Jeans' assumptions lead to linear acoustic momentum conservation equations for the gas, so the Jeans 1902 acoustic instability scale is $L_J = V_S / (\rho G)^{1/2}$, where V_S is the speed of sound, $\tau_g = (\rho G)^{-1/2}$ is the gravitational free fall time, ρ is the density and G is Newton's constant of gravitation.

The Jeans criterion for gravitational structure formation is that a large cloud of gas with density ρ is unstable to gravitational structure formation only on scales larger than J_H. By this criterion it is impossible to make any structures at all in the plasma epoch of the universe because the sound speed $V_{S-plasma} = c / 3^{1/2}$ is so large that $L_J > L_H$, where $L_H = ct$ is the horizon scale or scale of causal connection at time t after the big bang and c is the speed of light. Even after the plasma turns to gas at $t \approx 10^{13}$ seconds (300,000 years) the Jeans mass is an enormous million solar masses so the first stars cannot form until much later and planets can never form from gas at average temperatures because the density of the expanding universe decreases more rapidly than the sound speed with time. The problem with the Jeans criterion is that it is simply wrong [10]. Gravitational instability is absolute not linear [11], meaning that in a fluid with density fluctuations, structure will immediately begin to form at all scales of available density fluctuations unless prevented by the molecular diffusivity of the nearly collisionless non-baryonic dark matter or by viscous or turbulence forces. Mass moves toward density maxima and away from density minima at all scales unless prevented by forces or diffusivity.

When the plasma turns to gas it becomes a fog of planets [10] that become galaxy dark matter, as observed [20]. The only known form of non-baryonic dark matter NBDM is neutrinos, but there may be other massive neutrino-like particles formed in the early universe at temperatures unavailable in laboratories or there may be more ordinary neutrinos than known from present estimates. The evidence indicates the existence of something non-baryonic and quite massive that prevents the disintegration of galaxy clusters by gravitational forces, just as baryonic dark matter is needed to inhibit the disintegration of individual galaxies by centrifugal forces, but at this time no one knows what it is.

Density perturbations exist in the plasma because the big bang was triggered by a form of turbulent combustion at Planck temperature 10^{35} K termed the Planck-Kerr instability [12,13]. Planck particles and anti-particles appear by quantum tunneling and form the Planck scale equivalent of positronium. Positronium is produced by supernova temperatures of 10^{10} K sufficient to cause electron positron pair production, where the antiparticle pairs form a relatively stable combinations in orbit. Prograde captures give 42% release of the Planck particle rest mass energy which can only go into producing more Planck particles to fuel similar interactions. This gives a turbulent big bang with Reynolds number of about a million, terminated when cooling permits formation of quarks and gluons and a strong increase of viscosity and viscous stress that can cause an exponential increase in space to approximately a meter size in 10^{-33} seconds from 10^{-27} meters (10^{8} Planck lengths). The speed is about 10^{25} c. This inflation epoch produces the first fossil turbulence because all the turbulent temperature fluctuations of big bang turbulence are stretched beyond the scale of causal connection $L_H = ct$.

These fossils of big bang turbulence then trigger the formation of density fluctuations in the hydrogen and helium formed in the first three minutes by nucleosynthesis. The density fluctuations then seed the first gravitational structure formation [14]. The mechanism is not by the Jeans criterion but under the control of viscosity, turbulence or particle diffusivity, depending on which of three Schwarz scales is largest for the fluid in question. Because the collision length for non-baryonic dark matter is larger than $L_H = ct$ for the plasma epoch, only viscous forces or turbulence forces can prevent structure formation. Fragmentation of the non-baryonic dark matter occurs after the end of the plasma epoch, so this material is nearly irrelevant to the first structure formation, which begins when the horizon scale matches the Schwarz viscous scale at about 10^{12} seconds after the big bang (30,000 years) and the mass scale is that of superclusters. The plasma epoch is dominated by viscous forces with only weak turbulence because the photon viscosity is very large. Photons scatter from the free electrons of the plasma and transmit momentum because the electrons strongly couple to the ions to maintain electrical neutrality. Reynolds numbers are close to critical values so the

Schwarz viscous and turbulent scales are nearly identical. The first structures to form are voids because the universe is expanding. Condensations occur only after decoupling. The voids expand at the sonic speed of the plasma epoch which is $c / 3^{1/2}$, so we have significantly larger supervoids than superclusters as discussed in the following sections [15].

Because observations showed early structure must have occurred, cosmologists resorted to a *deus ex machina* solution (magic): they invented cold dark matter in order to create the present standard cosmological model. Following Jeans, it was assumed that a cold non-baryonic material must exist with sound speed sufficiently small that gravitational condensations in the plasma epoch could occur; that is, with $L_{J_{CDM}} < L_H$. Seeds of CDM were assumed to condense to provide gravitational potential wells into which the baryonic plasma would fall. The seeds would magically merge and stick over time, collecting more and more of the baryonic material until stars could form, and the clumps could cluster to larger and larger scales to produce galaxies and galaxy clusters and last superclusters of galaxies. This is hierarchical clustering, so the standard model is often called CDMHC. No CDM halo has ever been observed. In the following we suggest they do not exist and that CDM does not exist. Neither does the latest addition to the standard cosmological model; that is, dark energy and the cosmological constant Λ [16–19].

Instead, the plasma fragmentation beginning at 10^{12} s continues to smaller and smaller scales. Galaxies represent the smallest mass gravitational structures formed in the plasma before decoupling at 10^{13} s (300,000 years). Galaxy morphology reflects the turbulence existing in the plasma when they were formed. Protogalaxies fragmented along turbulent vortex lines at the Kolmogorov scale of the plasma and with a chain and spiral morphology demonstrated by the direct numerical simulations of Nomura. This morphology and a consistent protogalaxy length scale termed the Nomura scale is observed in a variety of Hubble Space Telescope observations discussed in the rest of this chapter [15].

At decoupling the kinematic viscosity ν decreased from photon viscosity values of $\sim 10^{26}$ m^2 s^{-1} to hot gas values of 10^{13} m^2 s^{-1}, a factor of ten trillion. This decreased the mass scale of fragmentation from that of protogalaxies to that of planets. Simultaneously the gas protogalaxies fragmented at the Jeans mass of globular star clusters, giving hot gas clouds of planet mass in million solar mass clumps. As the universe expands it cools. The gas clouds of hydrogen and helium eventually begin to freeze to form the galaxy dark matter, which is proto-globular-star-cluster PGC clumps of primordial-fog-particle PFP planets [10] as observed [20]. All stars form in PGCs from the trillion planets contained by a binary accretion process that gives larger and larger Jovian planets. The first stars to form were therefore small and long-lived, as observed in ancient globular star clusters. Structure formation from fluid mechanics naturally explains globular star clusters, which have always been a mystery. In standard ΛCDMHC cosmology, the first stars to form are superstars that re-ionize the universe to explain why so much hydrogen is missing in high red-shift quasar spectra. In our hydro-gravitational-dynamics HGD cosmology the missing hydrogen is tied up in frozen planets, the superstars (Population III) never happened, and Reionization of the Universe never happened [15].

Table 1. Length scales of gravitational instability

Length Scale Name	Definition	Physical Significance
Jeans Acoustic	$L_J = V_S/(\rho G)^{1/2}$	Acoustic time matches free fall time
Schwarz Viscous	$L_{SV} = (\gamma\nu/\rho G)^{1/2}$	Viscous forces match gravitational forces
Schwarz Turbulent	$L_{ST} = (\varepsilon/[\rho G]^{3/2})^{1/2}$	Turbulent forces match gravitational forces
Schwarz Diffusive	$L_{SD} = (D^2/\rho G)^{1/4}$	Diffusive speed matches free fall speed
Horizon, causal connection	$L_H = ct$	Range of possible gravitational interaction
Nomura, smallest plasma piece	$L_N = 10^{20}$ m	Proto-galaxy-fragmentation on vortex lines

V_S is sound speed, ρ is density, G is Newton's constant, γ is the rate of strain, ν is the kinematic viscosity, ε is the viscous dissipation rate, D is the diffusivity, *c* is light speed, *t* is time.

Theory

Figure 1 shows the sequence of gravitational structure formation events leading to the present time according to hydro-gravitational-dynamics HGD cosmology.

The big bang mechanism is one of turbulent combustion limited by gluon viscosity, where the maximum Taylor microscale Reynolds number $Re_\lambda = 1000$ occurred at $t = 10^{-33}$ seconds at 10^8 L_P, where $L_P = 10^{-35}$ is the Planck scale of quantum-gravitational instability [12,13]. The only particles possible during this epoch of high temperatures were Planck particles and Planck anti-particles that interact in the manner of electron-positron pair production to efficiently produce more Planck particles until the event cools enough for quarks and gluons to appear. Large negative Reynolds stresses expand space according to Einstein's theory of general relativity. During the big bang turbulence epoch, indicated in Figure 1a by the red star, the kinematic viscosity ν is the mean free path for particle collisions L_P times the speed of the particle. Even though the particle speeds were light speeds *c* at $t = 0$, the collision distances were small, giving a small initial kinematic viscosity $\nu_0 = 3 \times 10^{-27}$ m^2 s^{-1}. The Planck scale Reynolds number is near critical so L_P matches the Kolmogorov and Batchelor scales [3] of big bang turbulence and turbulent mixing.

The end of the big bang turbulence epoch occurred when the temperatures and particle speeds decreased as the Reynolds numbers increased till quarks and gluons could appear. This phase change is termed the strong force freeze out, with length scale $L_{SF} = 10^8$ L_P, or 10^{-27} m. Gluons are particles that carry momentum for the quarks, similar to photons carrying forces from momentum differences between the electrons and protons of the plasma epoch. It is assumed that the gluon viscosity is dramatically larger than the Planck viscosity, just as the photon viscosity is dramatically larger than the proton-proton collision viscosity of the plasma epoch, so that large negative gluon viscous stresses will accelerate the expansion of space exponentially during the inflation epoch indicated by the blue triangle. Other mechanisms such as the Guth false vacuum may be important for driving inflation. Whichever mechanism is dominant, many observations support an inflationary event.

During inflation, turbulent temperature fluctuations are fossilized by stretching beyond the scale of causal connection L_H. The mass-energy and entropy vastly increase, presumably by the rapid stretching of superstrings with Planck tension $c^4/G = 1.2 \times 10^{44}$ kg m s^{-2} [21]. The black dot in the inflation epoch symbolizes fossilized turbulent temperature fluctuations that can guide nucleosynthesis and preserve big bang turbulence information for testing using CMB temperature anisotropy patterns. In a period of 10^{-27} seconds the size of the big bang universe increases from 10^{-27} m to a few meters with density exceeding 10^{80} kg m^{-3}. The mass-energy of the universe within our present horizon scale $L_H = 10^{26}$ m is about 10^{52} kg, about 10^{-40} of the mass-energy created by the big bang. The mass-energy of the universe before inflation was less than 10^{16} kg so the increase was by a factor of 10^{74}. Guth describes this as the ultimate free lunch [21].

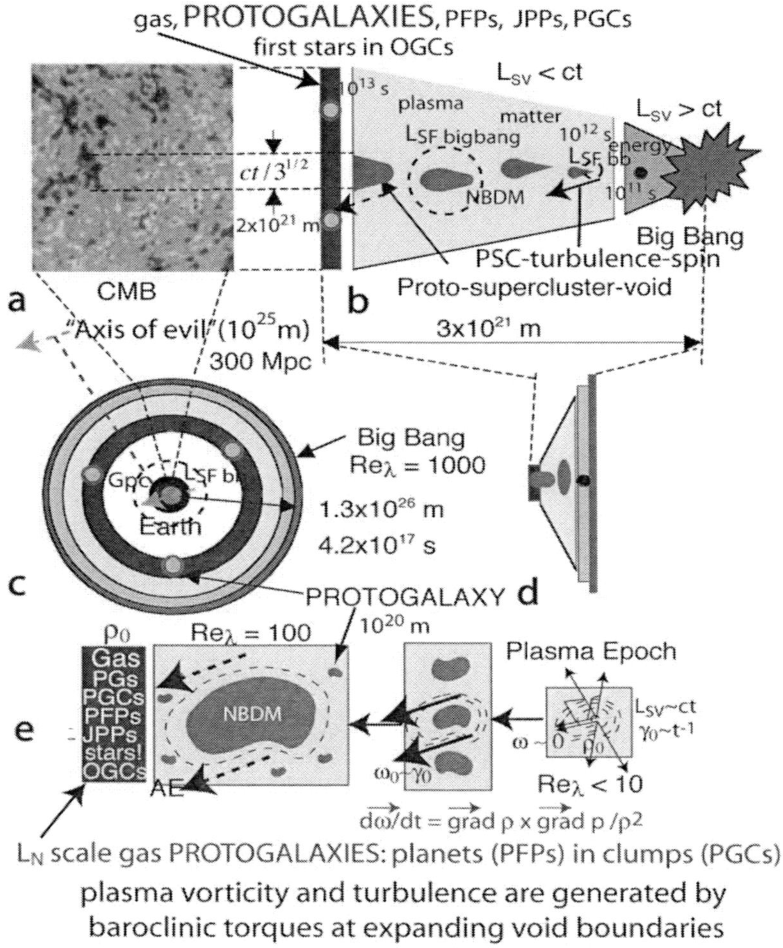

Figure 1. Protogalaxy formation at the end of the plasma epoch by hydro-gravitational-dynamics HGD theory. a. Cosmic Microwave Background temperature anisotropies reflect structures formed in the plasma epoch. b. From HGD the photon viscosity of the plasma epoch prevents turbulence until the viscous Schwarz scale L_{SV} becomes less than the Hubble scale (horizon scale, scale of causal connection) $L_H = ct$, where c is the speed of light and t is the time. The first plasma structures were proto-super-cluster voids and proto-super-clusters at 10^{12} seconds (30,000 years). c. Looking back in space is looking back in time. Protogalaxies were the last fragmentations of the plasma (orange circles with green halos) at 10^{13} seconds. d. The scale of the gravitational structure epoch is only $3x10^{21}$ m compared to present supercluster sizes of 10^{24} m and the largest observed supervoid scales of 10^{25} m. e. Turbulence in the plasma epoch is generated by baroclinic torques on the boundaries of the expanding super-voids.

The transition between energy domination and mass domination occurred at about 10^{11} s, the beginning of the plasma epoch shown in Figure 1b. The plasma is absolutely unstable and will fragment once the Schwarz viscous scale L_{SV} matches the horizon scale L_H. This occurs at time $t = 10^{12}$ s when the density of a flat universe is 10^{-15} kg m^{-3}. Most of this density, about 97%, is non-baryonic dark matter, NBDM, presumably some combination of neutrinos. The rest is plasma. None of it is dark energy. This gives a primordial density ρ_0 of 4×10^{-17} kg m^{-3}, which matches the density of old globular star clusters OGC. Because the universe continues to expand, fragmentation of the plasma at density minima is favored over condensation at density maxima. Voids appear in the plasma at 10^{12} seconds and proceed to expand as rarefaction waves at speeds limited by the speed of sound $c/3^{1/2}$ for 10^{13} seconds until decoupling. The material between the protosupercluster voids is protosuperclusters with mass 10^{46} kg, that of a thousand galaxies. The proto-super-clusters never collapse gravitationally but continue to expand with the universe. This is a key point; instead of a messy chaotic formation scenario of mergers the expansion was tranquil and preserved primordial alignments and densities. The present scale of superclusters is observed to be 10^{24} m.

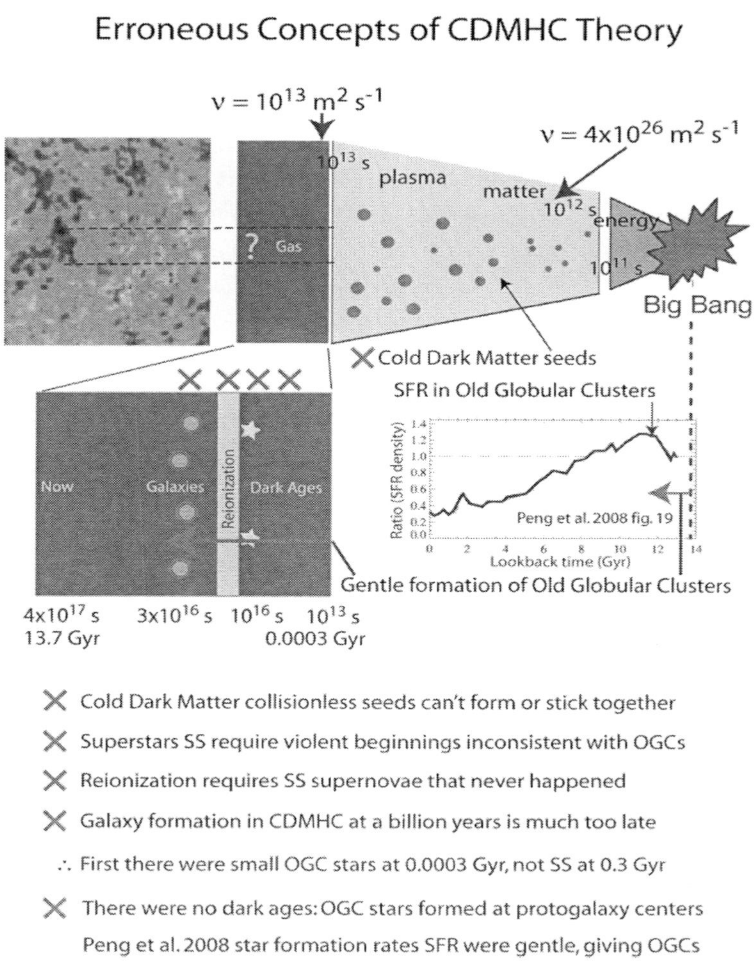

Figure 2. Galaxy formation by the standard ΛCDM model is impossible to reconcile with observations and fluid mechanical theory (see text). Failed aspects of the model are indicated by red Xs.

Smaller scale fragmentations occur in the plasma epoch at density minima up to the time of plasma to gas transition (decoupling). The last fragmentation is to produce protogalaxies at 10^{13} seconds, as shown in Figure 1b. The morphology of the protogalaxies is determined by weak turbulence generated by baroclinic torques on the surfaces of the supervoids as they expand. The direction of the spins is determined by the fossil strong force freeze-out density gradient at length scale L_{SF}. This spin direction determines the small wavenumber spherical harmonic directions for the CMB, the spins of galaxies in the local group, and even the spin of the Milky Way. All are directed toward the "axis of evil" [17], which has length scales in quasar polarizations to a Gpc, or 3×10^{25} m, as shown in Figure 1c.

Figure 2 illustrates erroneous concepts of cold dark matter hierarchical clustering (CDMHC) theory, which is at present the standard cosmological model along with the equally erroneous concept of dark energy and a cosmological constant Λ (ΛCDMHC). The standard model has no turbulence in the big bang but predicts Gaussian scale independent fluctuations and white noise spectrum of CMB fluctuations from the big bang and inflation epochs. All structure formation in the plasma epoch is attributed to the non-baryonic dark matter NBDM.

According to CDMHC seeds or halos of CDM form because their Jeans scale L_J is less than L_H. This is physically impossible because the diffusivity of any form of NBDM will cause the diffusive Schwarz scale L_{SD} of the material to exceed L_H. Even if a seed could form it could not stick to another seed because sticking requires collisions of the CDM particles, no matter how cold they might be. In Figure 2 each erroneous CDM concept is indicated by a red X.

It is claimed by CDMHC that the baryonic matter plasma is inviscid and falls into the CDM halos where it oscillates without viscous damping to give the acoustic peaks observed in the CMB temperature anisotropy spectrum. However, the plasma is far from inviscid, with a photon viscosity of $\nu = 4\times10^{26}$ m^2 s^{-1} [11]. Even if CDM halos existed, the plasma is far too viscous to enter the halo and oscillate acoustically.

Other erroneous concepts of CDMHC are listed in Figure 2. The CDM halos gradually cluster and collect gas for 300 My of dark ages (10^{16} s) in mini-galaxies which form the first superstars, one per mini-galaxy, in an enormous set of supernovae that re-ionizes all the gas formed at decoupling. The re-ionization concept is unnecessary to explain the lack of UV-absorbing gas as observed in quasar spectra because the gas is sequestered as dark matter planets in PGC clumps.

The first CDMHC galaxies to form occur much too late to be consistent with observations, at about a billion years after the big bang (3×10^{16} s). The star formation rate SFR in Peng et al. 2008 Figure 19 [23] cannot be superstars because superstars can only occur supported by violent turbulence stresses that would prevent formation of the very small stars observed to exist in old globular star clusters OGC.

From HGD and the mass of small stars it is easy to show from the Schwarz length scale criteria of Table 1 that the maximum viscous dissipation rate ε to permit the formation of a small OGC star is 10^{-12} m^2 s^{-3}. However, strong turbulence is required with ε values of 5×10^{-4} m^2 s^{-3} to permit formation of superstars and the re-ionization of the universe. If such large ε values existed before the first star formed, no old-globular-clusters OGCs could have formed because their small stars would be inhibited by turbulence produced as NBDM CDM halos filled with baryonic gas. Since OGCs are observed in all galaxies, the superstars and re-ionization concepts of CDMHC must be ruled out.

Observations

The fifth year WMAP observations of the CMB have been released, as shown in Figure 3. Figure 3a shows the Dunkley et al. 2008 5th year WMAP spherical harmonic temperature anisotropy data, with un-binned samples as light dots in the background [24].

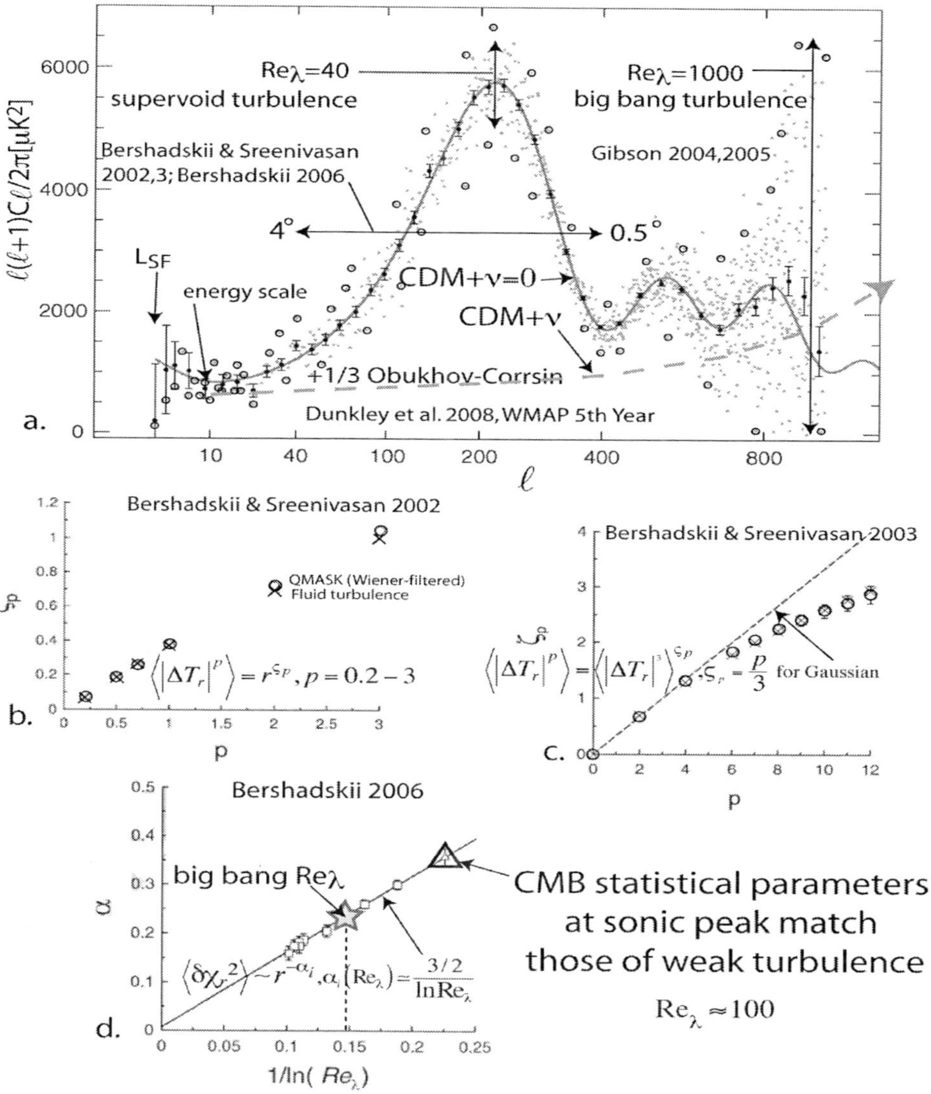

Figure 3. a. Spherical harmonic spectrum of CMB temperature anisotropies from the fifth year WMAP observations, Dunkley et al. 2008 [24]. The large sonic peak at wavenumbers near 200 reflects weak turbulence produced by gravitationally driven protosupervoid expansion according to Figure 1, not inviscid CDM (solid red). The dashed green line represents a viscous CDM model; that is, with no sonic signal and should match the +1/3 Obukhov-Corrsin turbulent mixing dissipation spectrum of big bang turbulence. Turbulence parameters are compared to CMB statistics in b. [25], c. [26] and d. [27] as described in the text.

The range of the un-binned samples in Figure 3a is a measure of the intermittency (non-Gaussianity) of the turbulence associated with the samples. The range indicated by circles and arrows is smaller at the sonic peak than at higher wavenumbers. This is attributed to the weak turbulence produced in the plasma at the boundaries of the expanding supervoids. The wider scatter at high wavenumbers k is attributed to the greater intermittency of big bang turbulence, which has a higher Reynolds number. An Obukhov-Corrsin turbulent mixing spectrum $\beta\chi\varepsilon^{-1/3}k^{-5/3}$ is fitted to the lowest wavenumber CMB spectrum, where β is a universal constant, χ is the dissipation rate of temperature variance and ε is the dissipation rate of velocity variance. It is assumed that the measured spherical harmonic CMB spectrum multiplied by $\ell(1+\ell)$ is not much different than a turbulent dissipation spectrum $k^2\phi$.

The red solid curve labeled CDM+ν=0 in Figure 3a is a CDM model without viscosity forced by ad hoc (GIGO) numerical assumptions to fit the data. Photon viscosity will not permit the baryonic plasma to enter the CDM potential wells even if they were physically possible, and will not permit any sonic oscillations because of viscous damping. Therefore the green dashed curve from big bang turbulence theory labeled CDM+ν should apply.

Evidence of turbulence from the expansion of super-voids producing baroclinic torques and vorticity at void boundaries is shown by a series of turbulence statistical parameters computed by Bershadskii and Sreenivasan from the low wavenumber CMB spectrum in Figure 3b, 3c, and 3d and compared to the same parameters computed for atmospheric, laboratory and numerical simulations of turbulence over a wide range of Reynolds numbers.

Figure 3b compares Weiner-filtered CMB data with extended self similarity turbulence data for various order structure functions. The agreement clearly demonstrates the CMB temperature anisotropies in the indicated angular size range 0.5 to 4 degrees studied were produced by turbulence. From HGD the turbulence reflects baroclinic torques at gravitationally driven expanding void boundaries.

Figure 3d shows an estimate of the sonic peak Reynolds number estimated by Bershadskii 2006 to be about $Re_\lambda \sim 100$. A more precise estimate is $Re_\lambda = 40$ as shown near the peak, using the expression derived for the temperature anisotropy variance as a function of separation distance r between sampling points. The power law exponent is a function of Re_λ. The value of Re_λ for the CMB is extracted using a series expansion of $\alpha(Re_\lambda)$ for high Reynolds number as a function of $1/\ln Re_\lambda$. It turns out for turbulence that only the first term is important, as shown, so that the CMB data near the sonic peak gives $Re_\lambda = 40$ as indicated by the bold triangle. For comparison, the big bang turbulence Re_λ value of 1000 is shown by a bold star. There can be little doubt from the combination of evidence shown in Figure 3 abcd that the plasma epoch was weakly turbulent in the range of scales including the sonic peak. As noted by Alexander Bershadskii (2001 personal communication to CHG) "the fingerprints of Kolmogorov are all over the sky".

As shown in Figure 1e, the last stage of plasma fragmentations guided by weak turbulence is the formation of proto-galaxy-mass objects with a linear morphology reflecting vortex tubes of the turbulence where the rates-of-strain that enhance fragmentation of the plasma to form voids is maximum.

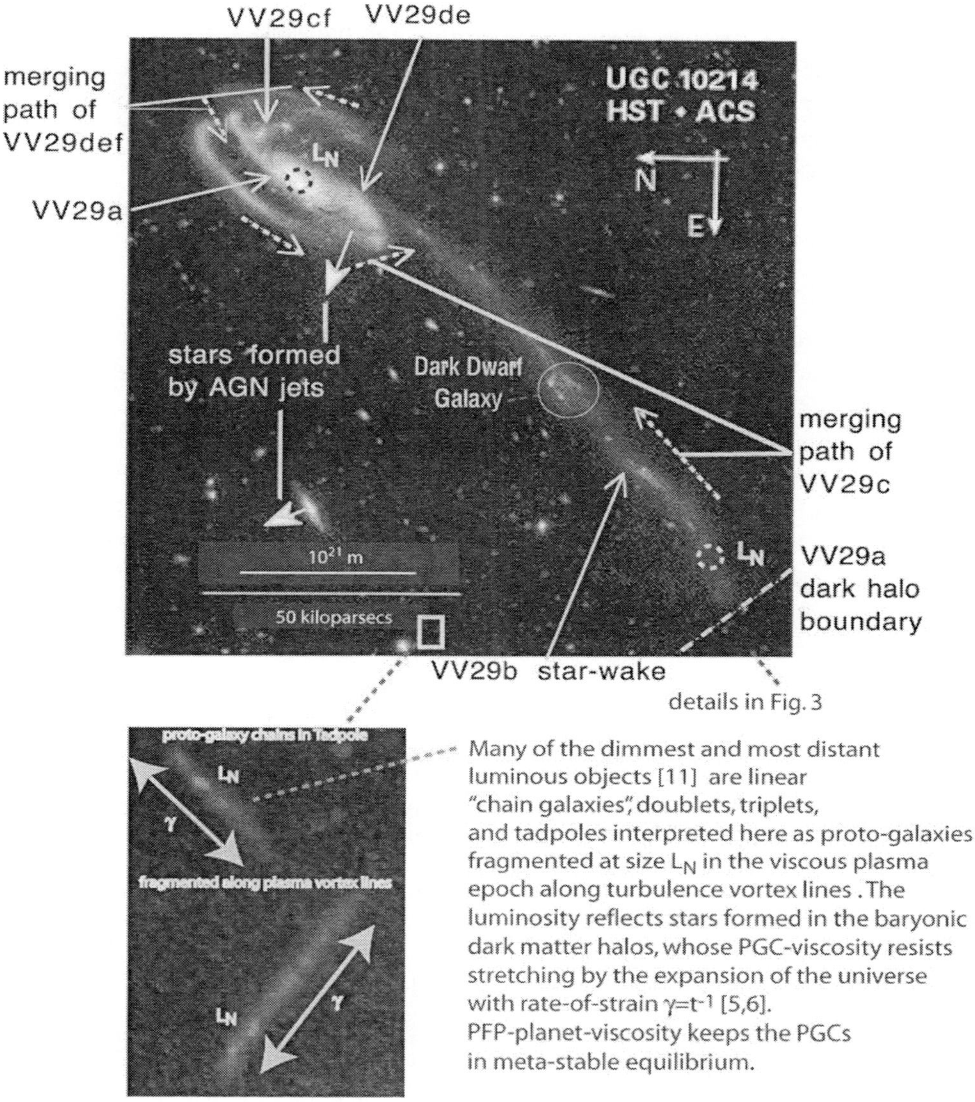

Figure 4. Tadpole galaxy merger (top), with background chains of protogalaxies shown in enlargement of small area at bottom center (bottom).

Figure 4 (bottom) is an image from the Hubble Space Telescope Advanced Camera for Surveys (HST-ACS) showing two proto-galaxy-chain clusters in the dim background of the Tadpole galaxy merger. Virtually all (94%) very dim galaxies from HST-ACS; that is, with magnitudes greater than 24, have such a linear morphology as expected from HGD [15]. The objects consist of bright nuclei separated by a dimmer glow that we interpret as proto-galaxies with a concentration of first-stars near the cores, with star-forming galaxy dark matter (planets in clumps) in between.

Figure 4 (top) shows the formation of star trails and dust trails in the baryonic dark matter halo of the Tadpole central galaxy VV29a as galactic objects VV29cdef merge in frictional spirals on the VV29a disk plane [19]. A sharp VV29a dark halo boundary is clear in the high-resolution HST-ACS images. The eponymous VV29b spiral filament shows numerous young globular star clusters [28] aligned in a direction pointing precisely at the point of frictional merger as the objects move through the dark matter halo triggering star formation by tidal forces in a L_N scale diameter star wake. The size of VV29a dark matter halo diffused from the L_N scale central core of old small stars is $8x10^{21}$ m, or 0.9 Mpc. From this evidence, the concept of frictionless tidal tails in galaxy mergers appears to be obsolete [29].

Conclusion

A fluid mechanical analysis of gravitational structure formation termed hydro-gravitational-dynamics HGD shows that the CDMHC model is inconsistent with theory and observations and must be abandoned. From HGD, gravitational structure formation is driven entirely by the baryonic matter of the plasma epoch. The first structures are proto-super-cluster-voids and proto-super-clusters, followed by proto-galaxies at decoupling. The proto-galaxies are formed in very weakly turbulent plasma, but possess the morphology of weak turbulence in that the protogalaxies are stretched along the vortex lines of the plasma turbulence with a diameter reflecting their viscous-inertial-vortex-gravitational origin at the Nomura scale 10^{20} m. Although our description of HGD seems to emphasize turbulence processes, the universe described is in fact more tranquil than previously envisioned, and can preserve primordial alignments seen in many cosmological structures, in particular proto-galaxies.

References

[1] Darwin, G. H. 1889. On the mechanical conditions of a swarm of meteorites, and on theories of cosmogony, *Phil. Trans.*, **180**, 1-69.

[2] Jeans, J. H. 1902. The stability of spherical nebula, *Phil. Trans.*, **199A**, 0-49.

[3] Gibson, C.H. (1991). Kolmogorov similarity hypotheses for scalar fields: sampling intermittent turbulent mixing in the ocean and galaxy, *Proc. Roy. Soc. Lond.* A, **434**, 149-164.

[4] Gibson, C. H. (2006). *Turbulence, update of article in Encyclopedia of Physics*, R. G. Lerner and G. L. Trigg, Eds., Addison-Wesley Publishing Co., Inc., pp.1310-1314.

[5] Gibson, C. H. (1981). Buoyancy effects in turbulent mixing: Sampling turbulence in the stratified ocean, *AIAA J.*, **19**, 1394.

[6] Gibson, C. H. (1968a). Fine structure of scalar fields mixed by turbulence: I. Zero-gradient points and minimal gradient surfaces, *Phys. Fluids*, **11**: 11, 2305-2315.

[7] Gibson, C. H. (1968b). Fine structure of scalar fields mixed by turbulence: II. Spectral theory, *Phys. Fluids*, **11**: 11, 2316-2327.

[8] Gibson, C. H. (1986). Internal waves, fossil turbulence, and composite ocean microstructure spectra, *J. Fluid Mech.* **168**, 89-117.

[9] Gibson, C. H. (1999). Fossil turbulence revisited, *J. of Mar. Syst.*, **21**(1-4), 147-167, astro-ph/9904237.

[10] Gibson, C.H. (1996). Turbulence in the ocean, atmosphere, galaxy and universe, *Appl. Mech. Rev.*, **49**, no. 5, 299-315.

[11] Gibson, C.H. (2000). Turbulent mixing, diffusion and gravity in the formation of cosmological structures: The fluid mechanics of dark matter, *J. Fluids Eng.*, **122**, 830–835.

[12] Gibson, C.H. (2004). The first turbulence and the first fossil turbulence, *Flow, Turbulence and Combustion*, **72**, 161–179.

[13] Gibson, C.H. (2005). The first turbulent combustion, Combust. *Sci. and Tech.*, **177**: 1049–1071, arXiv:astro-ph/0501416.

[14] Gibson, C.H. (2006). *The fluid mechanics of gravitational structure formation,* astro-ph/0610628.

[15] Gibson, C.H. (2008). Cold dark matter cosmology conflicts with fluid mechanics and observations, *J. Applied Fluid Mech.*, Vol. 1, No. 2, pp 1-8, 2008, arXiv:astro-ph/0606073.

[16] Gibson, C.H. & Schild, R.E. (2007*). Interpretation of the Helix Planetary Nebula using Hydro-Gravitational-Dynamics: Planets and Dark Energy*, arXiv:astro-ph/0701474.

[17] Schild, R.E & Gibson, C.H. (2008). *Lessons from the Axis of Evil*, axXiv[astro-ph]:0802.3229v2.

[18] Gibson, C.H. & Schild, R.E. (2007). *Interpretation of the Stephan Quintet Galaxy Cluster using Hydro-Gravitational-Dynamics*: Viscosity and Fragmentation, arXiv[astro-ph]:0710.5449.

[19] Gibson, C.H. & Schild, R.E. (2002). *Interpretation of the Tadpole VV29 Merging Galaxy System using Hydro-Gravitational Theory*, arXiv:astro-ph/0210583.

[20] Schild, R. 1996. Microlensing variability of the gravitationally lensed quasar Q0957+561 A,B, *Ap J*, **464**, 125.

[21] Greene, B. 1999. *The Elegant Universe, Superstrings, Hidden Dimensions, and the Quest for the Ultimate Theory*, W. W. Norton & Compant, New York.

[22] Guth, A. H. 1997. *The Inflationary Universe, The Quest for a New Theory of Cosmic Origins*, Helix Books, Addison-Wesley Pub. Co., Inc., New York.

[23] Peng, E. W. et al. 2008, The ACS Virgo Cluster Survey. XV. The Formation Efficiencies of Globular Clusters in Early-Type Galaxies: The Effects of Mass and Environment, *Ap J,* **681**, 197-224.

[24] Dunkley, J. et al. (2008). Five-year Wilkinson Microwave Anisotropy Probe (WMAP) Observations: Likelihoods and Parameters from the WMAP data, subm. *Ap JS;* arXiv:0803.0586v1.

[25] Bershadskii, A., and Sreenivasan, K.R. 2002. Multiscaling of cosmic microwave background radiation, *Phys. Lett. A,* **299**, 149-152.

[26] Bershadskii, A., and Sreenivasan, K.R. 2003. Extended self-similarity of the small-scale cosmic microwave background anisotropy. *Phys. Lett. A*, **319**, 21-23.

[27] Bershadskii, A. 2006. Isotherms clustering in cosmic microwave background, *Physics Letters A*, **360**, 210-216.

[28] Tran, H. D., Sirianni, M., & 32 others 2003. Advanced Camera for Surveys Observations of Young Star Clusters in the Interacting Galaxy UGC 10214, *Ap J*, **585**, 750.

[29] Toomre, A., & Toomre, J. 1972. Galactic Bridges and Tails, *Ap J,* **178**, 623.

INDEX

2

A

B

C

H

I

L

M

T

U

V

W

Y